Cell Biology

MODERN VIEWS IN BIOLOGY 🍁

Volumes in preparation:

REPRODUCTIVE BIOLOGY
NEUROBIOLOGY AND BEHAVIOUR

Cover photographs The colour image is a computer graphics model of a crystalline insulin hexamer, as found in storage granules: the image was produced by the Laboratory of Molecular Biology, Department of Crystallography, Birkbeck College, University of London. The black and white photograph is an electron micrograph through a section of a mammalian pancreas ($\times$ c. 26 000): micrograph courtesy of Dr S. Howell, Department of Physiology, Queen Elizabeth College, University of London.

Cell Biology

EDITED BY

Dr. B. King

Department of Biology, Stowe School, Buckingham, England

London
ALLEN & UNWIN
Boston   Sydney

Allen & Unwin (Publishers) Ltd,
40 Museum Street, London WC1A 1LU, UK

Allen & Unwin (Publishers) Ltd,
Park Lane, Hemel Hempstead, Herts HP2 4TE, UK

Allen & Unwin Inc.,
8 Winchester Place, Winchester, Mass. 01890, USA

Allen & Unwin (Australia) Ltd,
8 Napier Street, North Sydney, NSW 2060, Australia

First published in 1986

British Library Cataloguing in Publication Data

King, B.
   Cell biology.—(Modern views in biology; v. 1)
1. Cells
I. Title   II. Series
574.87   QH581.2
ISBN 0-04-574026-7

Library of Congress Cataloging in Publication Data

   Cell biology.
(Modern views in biology)
Includes bibliographies and index.
1. Cytology.   I. King, B. (Barry), 1947-
II. Series.   [DNLM: 1. Cells. QH 581.2 C3915]
QH581.2.C42   1985   574.87   85-15093
ISBN 0-04-574026-7 (pbk.: alk. paper)

Set in 9 on 11 point Melior by Columns of Reading
and printed in Great Britain by Hazell Watson and Viney Ltd,
Member of the BPCC Group, Aylesbury, Bucks

Editor's foreword

The student of biological sciences at Advanced level or in the first year of an undergraduate course is expected to be aware of the changing nature of biology. New research is published in a variety of specialised journals and even if these journals are available they are likely only to confuse students. Review articles often assume a detailed background knowledge of the subject and hence are of limited value. Such articles are written with little knowledge of the information contained in most Advanced level textbooks. In addition, it is no longer possible for a single text to cover the whole field of biology while remaining up to date. Teachers and students at school, college or university who try to keep abreast with advances are thus likely to be faced with a perplexing and often conflicting array of information. The aim of 'Modern Views in Biology' is to review the information contained in Advanced level texts while presenting more current ideas in the fields where significant advances have been made.

The difficulty of dividing a course into a number of volumes is vast and opinion will undoubtedly differ on how it is best done. In devising 'Modern Views in Biology' the amount of overlap has been reduced to a minimum. Each volume consists of a number of articles on closely related topics and can thus be used in isolation or in any sequence with the others. Collectively, the volumes provide an up-to-date coverage of the biology studied at Advanced level and thus are likely to be of tremendous value to teachers and students.

Articles have been written by research workers in association with an experienced Advanced level teacher. In this way the current state of knowledge in a particular field is reviewed while the basic conceptual framework encountered in Advanced level textbooks is not forgotten. In addition, authors have been asked to outline the likely direction of future research together with any possible applied aspects which may result from such work. A concerted effort has been made by authors and editor to maintain a constant level of presentation. However, each author has been encouraged to approach his own subject in his own way in order that the author's enthusiasm might be communicated to the reader.

Barry King

Acknowledgements

We are grateful to the following for permission to reproduce copyright material:

Figure 2.1a reproduced from E. K. Neale and G. B. Chapman, *Journal of Bacteriology* **104** 525 (1970), by permission of the American Society for Microbiology; Figure 2.1b reproduced from J. D. Dodge and R. M. Crawford, in M. A. Sleigh (ed.), *The biology of protozoa* (1973), by permission of J. D. Dodge and Edward Arnold; Cancer Research Campaign (4.11 & 19); Figure 5.5 reproduced from J. N. Davidson, *Biochemistry of Nucleic Acids*, Chapman & Hall; Figures 5.11a and 6.5 redrawn from J. D. Watson, *Molecular Biology of the Gene*, © 1976 The Benjamin/Cummings Publishing Company; Figure 5.11b reproduced from J. Cairns, *Cold Spring Harbor Symp. Quant. Biol.* **28** 44 (1963); Figure 5.16 from J. G. Gall and E. H. Atherton, *Cold Spring Harbor Symp. Quant. Biol.* **38** (1974); Figure 5.18 reproduced from *The Journal of Cell Biology*, 1975, **64** 532 by copyright permission of the Rockefeller University; Garland Publishing (5.20 & 25); Table 5.1 from Srb and Horowitz, *J. Biol. Chem.* **154**, 133 (1944); John Wiley (Tables 5.3 & 4); Figure 6.1 redrawn from *The control of gene expression in animal development* (J. B. Gurdon 1974) by permission of Oxford University Press; W. H. Freeman (6.33).

Contents

List of tables

List of contributors

M. R. Hartley, Department of Biology, University of Warwick, Coventry,
   CV4 7AL, UK
G. E. Jones, Department of Biology, Queen Elizabeth and Kings College,
   London, W8 7AH, UK
B. King, Biology Department, Stowe School, Buckingham, MK18 5EH, UK
M. A. Sleigh, Department of Biology, University of Southampton,
   Southampton, SO9 3TU, UK
C. H. Wynn, Department of Biochemistry, University of Manchester,
   Manchester, M13 9PL, UK

Abbreviations

A	adenine
ADP	adenosine diphosphate
Ala	alanine
AMP	adenosine monophospate
cAMP	cyclic AMP
Arg	arginine
Asn	asparagine
Asp	aspartate
ATP	adenosine triphosphate
Butyl-PBD	(2-(4-t-butylphenyl)-5-(4″-diphenylyl)-1,3,4-oxadiazole)
C	cytosine
CAP	catabolite gene activator protein diphosphate
CDP	cytosine diphospate
CMP	cytosine monophosphate
CTP	cytosine triphosphate
CoA	coenzyme A
cyclic AMP	adenosine 3′, 5′-cyclic monophosphate
Cys	cysteine
d	2′-deoxyribo
DNA	deoxyribonucleic acid
cDNA	complementary DNA
EM	electron microscope
ER	endoplasmic reticulum
FAD	flavin adenine dinucleotide (oxidised)
FADH$_2$	flavin adenine dinucleotide (reduced)
fMet	formylmethionine
G	guanine
Gln	glutamine
Glu	glutamate
Gly	glycine
GDP	guanosine disphosphate
GMP	guanosine monophosphate
GTP	guanosine triphosphate
His	histidine
HLA	histocompatibility antigen
hnRNA	heterogeneous nuclear RNA
Ile	isoleucine
kdal	a unit of mass equal to 1000 dalton, the terms dalton and molecular weight are interchangeable
Leu	leucine
LH	luteinizing hormone
Lys	lysine
MAP	Microtubule-associated protein
Met	methionine
mRNP	messenger ribonuclearprotein
MTOC	microtubule organising centre

NAD+	nicotinamide adenine dinucleotide (oxidised)
NADH	nicotinamide adenine dinucleotide (reduced)
NADP	nicotinamide adenine dinucleotide phosphate (oxidised)
NADPH	nicotinamide adenine dinucleotide phosphate (reduced)
NTP	nucleoside triphosphate
Phe	Phenylalanine
Pi	for red light absorbing form of phytochrome
PPi	inorganic pyrophosphate
Pfr	inorganic orthophosphate
PPo	2, 5-diphenyloxazole
Pr	red light absorbing form of phytochrome
Pro	proline
RER	rough endoplasmic reticulum
RNA	ribonucleic acid
mRNA	messenger RNA
rRNA	ribosomal RNA
tRNA	transfer RNA
Ser	serine
T	thymine
Thr	threonine
TSH	thyroid stimulating hormone
Trp	tryptophan
Tyr	tyrosine
U	uracil
UDP	uridine diphosphate
UDP-glucose	uridine diphosphate glucose
UMP	uridine monophosphate
UTP	uridine triphosphate
Val	valine

Cell Biology

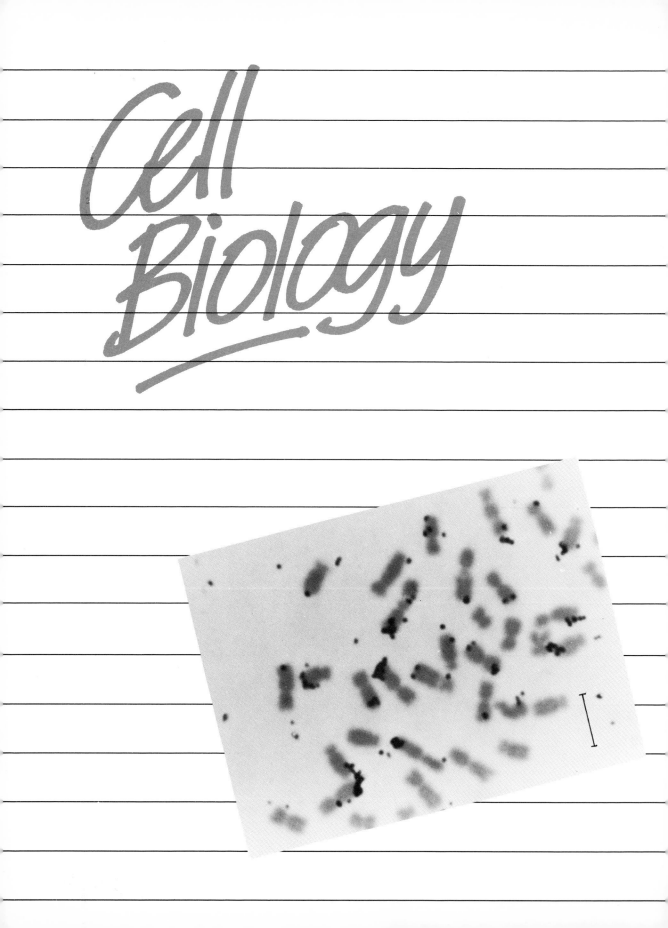

1

Techniques in cell biology

B. King

MODERN VIEWS IN BIOLOGY

1.1   Introduction

During the past few decades there has been a tremendous growth in the biological sciences. There are many factors underlying this increase in our knowledge of living systems. The development of new techniques in the physical and chemical sciences has been important. It has been argued that the subsequent application of these methods in biology is the single most important factor in the increase of knowledge we have witnessed in the last 30 years.

This chapter is not intended to provide students with a comprehensive treatment of the techniques used in cell biology. It is designed to give the reader an understanding of the methods discussed in the following chapters of this volume.

1.2   Light microscopy

1.2.1   MAGNIFICATION AND RESOLUTION

The oldest and still the most widely used instrument for studying the structure of organisms and cells is the light microscope. Students of biology often regard the microscope as no more than a magnifying device and do not really appreciate *the vital property of resolution*. **Magnification** achieved by microscopy can be defined as the ratio of the apparent size of the object to the actual size of the structure. In practice, this can be calculated by multiplying the primary magnification of the objective lens by that of the eyepiece. It is possible to increase the magnifying power of the instrument by increasing the power of the objective or eyepiece or both. Above a certain level, simple enlargement *does not increase the amount of detail but only serves to increase the size of the image*.

The recognition of individual but closely spaced points is termed **resolution**. Each optical instrument has a maximum resolving power; for the eye it is reached when two points come to lie approximately 80 μm apart. Below this distance the eye will just see one point: use of a magnifying glass or microscope would, of course, resolve the points.

The absolute resolving power of the light microscope *is limited by the wave nature of light itself*. A single light ray is scattered by the specimen, the smaller the feature the greater the angle of scatter. The angle of rays gathered by the objective is important in the determination of resolution. This angle is expressed in terms of **numerical aperture**, NA (Fig. 1.1b):

$$NA = n \sin u$$

where n is the refractive index of the medium between the specimen and the lens, and u is half the angle of light accepted by the objective.

Resolution is given approximately by the formula

$$\text{resolution} = \frac{0.61\lambda}{NA}$$

where λ is the wavelength of light.

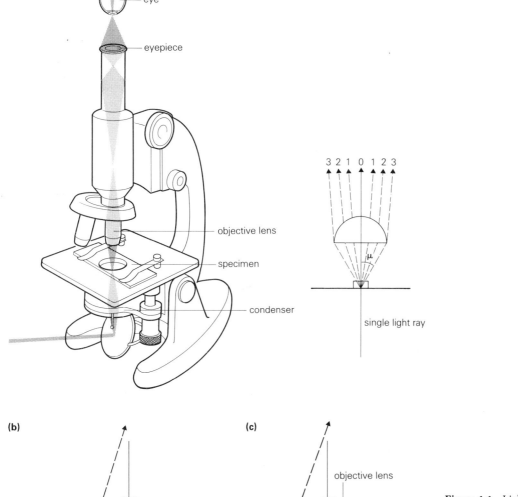

(a)

eye

eyepiece

objective lens

specimen

condenser

3 2 1 0 1 2 3

μ

single light ray

(b)

objective lens

air

coverglass

$u = 39°$

(c)

objective lens

immersion oil

coverglass

$u = 60°$

Figure 1.1 Light microscopy: (a) conventional ray pattern in a light microscope together with a single ray; (b) dry objective with a numerical aperture of 0.94; (c) oil immersion objective, numerical aperture increased to approximately 1.2.

As light travels through objects its speed may be slowed according to the refractive index of the medium. The use of immersion oil, which has an index of 1.5 (similar to that of glass), increases the numerical aperture of the objective lens (Fig. 1.1c). In addition, the distance between the specimen and the objective is reduced. In simple terms these two changes are effective in allowing a greater degree of information to be collected from the specimen, and hence a higher resolving power can be achieved.

1.2.2 STAINING

Direct observation of cells reveals a certain degree of information. However, the relatively transparent nature of cytoplasm means that there is little effect on light passing through the cells or tissues. The use of specific staining techniques greatly increases the usefulness of any microscope. In recent years, developments in the field of histochemistry have enabled scientists to stain specific molecules and hence cell organelles.

The staining of deoxyribonucleic acid (DNA) in cells by the technique of Feuglen and Rossenbeck is an example with which students may be familiar. The DNA in the nuclear material is treated with warm HCl. This causes partial hydrolysis, producing *deoxyribose components with exposed aldehyde groups*. These in turn react with Schiff's reagent, producing an insoluble purple compound.

Many of the stains used are toxic to living cells and hence only effective with dead tissue. Fortunately, there is another group of stains which can be incorporated into living cells without dramatically affecting the necessary functions of the tissue. These are termed **vital stains**.

In addition to staining techniques, a number of other methods that allow the examination of specimens have been developed. These can necessitate optical modifications to the microscope and some of the more important are described in the next section.

1.3 Contrast techniques

1.3.1 DARK GROUND MICROSCOPY

Although this method is not commonly used in schools it is an extremely valuable technique for studying certain types of cells and tissue. The optical system of the microscope is virtually reversed, *a bright image appearing against an essentially dark background*. A hollow cone of light is created as shown in Figure 1.2 by placing a central circular stop below the condenser.

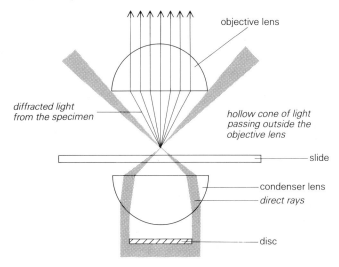

Figure 1.2 Optical arrangement used in dark ground illumination. The insertion of an opaque disc below the condenser prevents any of the direct light from the specimen entering the objective lens.

No direct light rays are gathered by the objective lens; the light rays diffracted from the specimen create a bright image on a black background. The use of dark ground illumination enables examination of details of the structure of aquatic organisms such as protozoa, rotifers and small crustaceans.

1.3.2 PHASE CONTRAST

In order to appreciate the usefulness of the technique of phase contrast a certain knowledge of the wave nature of light is required. The electromagnetic theory considers light to be associated with variations in electric and magnetic fields. This idea is illustrated in Figure 1.3a, together with several of the important parameters of light. The amplitude of a light wave is important as it determines the intensity of the light: the intensity is proportional to the square of the amplitude. The colour of the light is determined by its wavelength (λ). Light with a wavelength of 450 nm is detected as blue while a wavelength of around 660 nm gives red light.

Another important property of light is its **phase**. If two light waves are completely in phase (Fig. 1.3b), interference occurs between the waves; the resultant amplitude is greater and a much brighter light is seen. At the other extreme, if two waves are out of phase, the resultant wave has an amplitude of zero (Fig. 1.3c). In this case nothing is seen. When cells are viewed using normal light optics with no staining techniques, the indirect light waves passing through the cytoplasm are likely to be retarded. This amounts to around $\lambda/4$ relative to the direct rays (Fig. 1.3d) passing through the aqueous medium. Although interference has occurred it is usually insufficient to enable the eye to detect the detail within the cell, and the cytoplasm appears transparent.

The principle behind phase contrast microscopy lies in the *further retardation of these indirect waves*. If the difference between these two different waves can be increased to approximately $\lambda/2$ then the interference created between the two sets of waves will be sufficient to reveal details of the cytoplasm. This is achieved by the insertion of a phase plate within the objective lens (Fig. 1.4a). The plate is a glass disc in which an annular groove has been cut. The light waves which have been diffracted by the specimen are forced through the central region of the phase plate (Fig. 1.4b). This area is thicker and retards these light rays by $\lambda/4$. Since the waves are already out of phase with the direct rays, a total phase shift of $\lambda/2$ is achieved. This phase shift is ensured by placing an annular diaphragm before the condenser, which forces the majority of the direct light waves through the inner part of the phase plate.

Since the separation of direct and diffracted light rays can never be perfect, a small proportion of the diffracted light will in fact pass through the groove in the phase plate. This leads to a 'halo' appearing around the object. This effect can be seen quite clearly in Figure 7.15a. The use of phase microscopy has enabled such processes as cell division and endocytosis to be studied in living cells.

1.3.3 FLUORESCENCE MICROSCOPY

Certain compounds absorb short wavelength radiation and then re-emit energy as light of a longer wavelength. This phenomenon is known as **fluorescence** and is now routinely used in microscopy. The fluorescent microscope (Fig. 1.5) is essentially an ordinary optical instrument which has

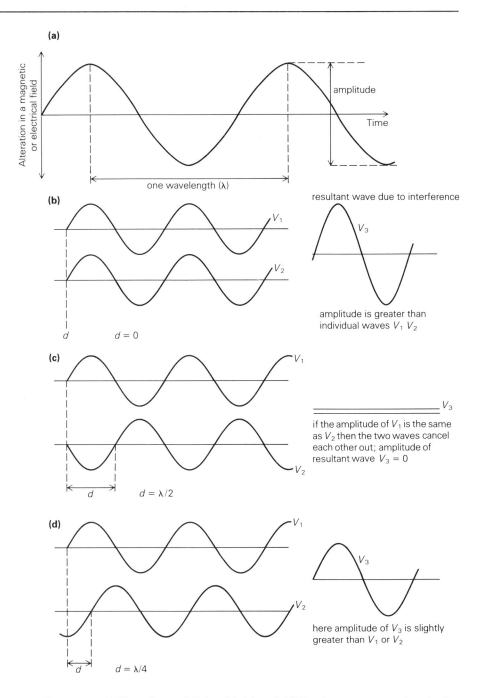

Figure 1.3 (a) Wave form of light. (b), (c) and (d) Resultant waves produced after interference between waves. The degree to which the waves are cut of phase (d) affects the amplitude of the resultant wave.

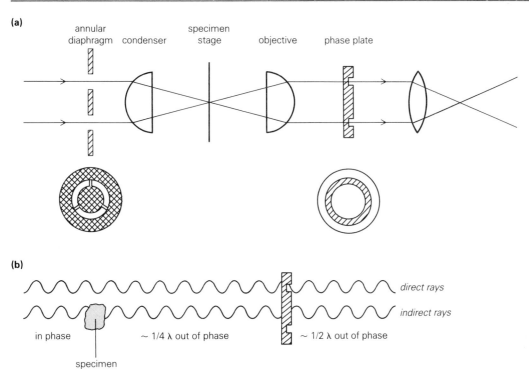

Figure 1.4 (a) Optical system of a phase contrast microscope. (b) Enhanced contrast ensured by the phase plate.

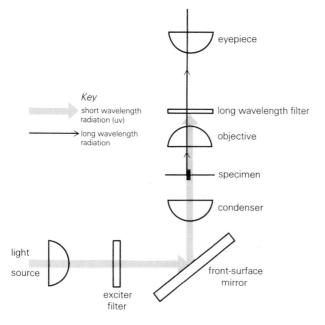

Figure 1.5 Fluorescent light microscopy. The pathway of the short wavelength light is shown by the thick line. Longer wavelength fluorescence produced by the specimen is shown by the thinner line.

been modified by incorporation of special filters. The light source is usually a high-intensity lamp which emits radiation in the blue-violet or ultra-violet region of the spectrum. This radiation passes through an 'exciter' filter which allows only the required wavelengths to pass. This light is capable of causing fluorescence in the specimen. The placing of a second filter beyond the objective lens removes any of the short exciting radiation while allowing the longer wavelength fluorescence to form an image. Since the image is formed entirely by light emanating from the specimen, fluorescing objects appear as *intensely bright images in a uniformly dark background.*

Compounds which naturally fluoresce exist in most biological tissues, for instance collagen and chlorophyll. It is also possible to induce fluorescence in certain compounds by chemical means. Amines can be modified by formaldehyde treatment so that they then exhibit fluorescence. This has enabled substances such as dopa, dopamine and andrenalin to be located within nerve cells. Fluorescent dyes '**fluorochromes**' are now available and these can be introduced into cells. An important application of such dyes has been in the investigation of chromosome behaviour.

The most sophisticated application of fluorescent microscopy so far

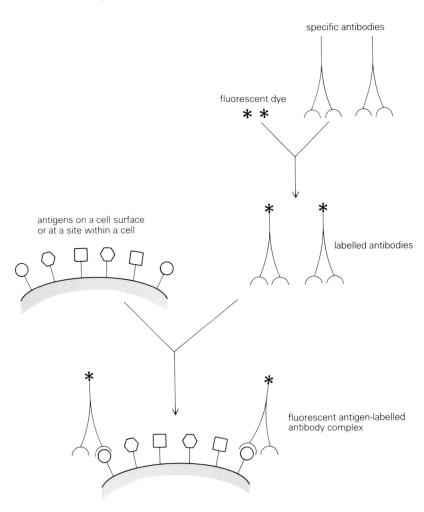

Figure 1.6 Stages in the technique of fluorescent antibody labelling.

developed has been in the field of **immuno-fluorescent antibody labelling**. Antibodies can be prepared against particular cell proteins. The attachment of a fluorochrome to an antibody creates a very powerful analytical tool (Fig. 1.6): using this method the exact location of certain cellular proteins has now been found. One of the most dramatic discoveries using this technique has been that of the **cytoskeleton** (see §7.3 & Fig. 7.10). Antibodies to actin are covalently labelled with a fluorescent molecule such as fluoresein. The addition of this complex to cells in culture followed by fluorescent microscopy reveals a highly ordered array of filaments. Using similar techniques the cellular location of the protein vinculin is revealed in Figure 7.17.

1.4   Electron microscopy

Although the light microscope with all its variation and modifications remains a powerful research tool, the fundamental wave properties of light impose a *limitation on the resolving power* of the instrument. The development of the electron microscope (EM) enabled a much greater resolving power to be achieved. A beam of electrons exhibits many of the properties similar to those shown by light waves. In particular, the electrons have a wavelength that is dependent upon their energy. This wavelength means a resolution of 0.5 nm becomes theoretically possible. In fact this resolution remains unattainable for a variety of reasons, one of the most important being the relatively inefficient nature of the magnetic lenses needed to focus the electron beams.

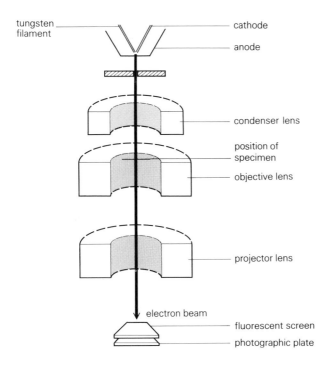

tungsten filament — cathode
anode
condenser lens
position of specimen
objective lens
projector lens
electron beam
fluorescent screen
photographic plate

Figure 1.7   The transmission electron microscope.

The source of electrons is a heated tungsten filament. These electrons are then accelerated towards a positively charged anode (Fig. 1.7). The application of a potential difference of up to 100 kV produces a beam of electrons with an extremely short wavelength. The beam passes through a hole in the centre of the anode and then follows a pathway similar to that in a conventional optical microscope (see Fig. 1.1). The whole of the electron microscope column operates at a high vacuum which prevents collisions between the electrons and gas molecules. Such collisions would scatter the beam, so reducing the number of electrons which might pass through the specimen and eventually contribute to an image. However, the vacuum means that *living cells cannot be observed in a conventional EM*. This problem can be at least partly solved by the use of a high voltage EM (see §1.4.5).

The poor penetrative power of electrons also means that thin specimens must be prepared for the EM. The standard procedure for the preparation of specimens entails fixation, dehydration, staining and sectioning. This is similar to the techniques used in the preparation of sections for the light microscope, the most significant difference being the need for ultra-thin sections. Depending upon the tissue and type of investigation being performed, the sections can be between 10 and 100 nm. To obtain such sections, the tissue blocks are sectioned on an **ultramicrotome** using a glass or diamond knife. Typical examples of electron micrographs are shown in Figures 2.1 and 7.2.

1.4.1   STAINING TECHNIQUES

As has already been mentioned, electrons are scattered by collisions with atoms. The extent to which scattering occurs depends upon the size and concentration of atoms encountered. The larger the atom (higher the atomic number), the greater the probability of collisions. The elements present in biological material are relatively light, e.g. hydrogen, carbon, nitrogen, phosphorus and sulphur. In order to differentiate between regions of the cell it is often necessary to introduce heavy metal stains into the section.

1.4.2   NEGATIVE STAINING

Although microscopy is usually thought to be concerned with the investigation of the internal organisation of a structure, very often the shape and surface features are of importance. This is clearly true in the case of viruses and isolated macromolecules. In the technique of negative staining a heavy metal stain is allowed to cover the specimen (Fig. 1.8a). The stain accumulates around the object and adheres to it on drying. The electron beam will be scattered by the stain while passing largely unaffected through the specimen. This produces a light object on a dark background.

1.4.3   SHADOWING

This technique is really an adaptation of negative staining. Once again heavy metal stains are used. However, here a metallic element is selected rather than a metal salt. The element is evaporated by high-temperature treatment and the metal atoms allowed to coat the specimen. If the stream of evaporated metal hits the object at an angle the stain will not be deposited in the 'shadow' of the specimen (Fig. 1.8b). The technique is commonly used to reveal details of the external structure of an object. In addition, it is possible to calculate the dimensions of a structure if the *angle of shadowing is known*: the length of the shadow is directly related to the height of the object.

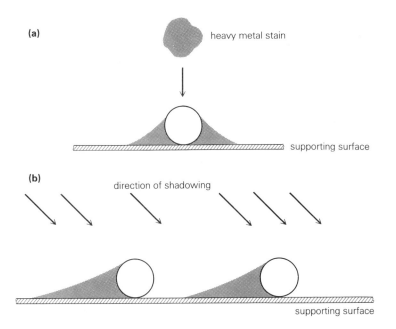

Figure 1.8 (a) Negative staining: the specimen appears light against a dark background. (b) Shadowing technique: evaporated metal is deposited across the specimen.

1.4.4   FREEZE FRACTURE

In recent years the technique of freeze fracture has added a great deal of information about the internal structure of cells. In particular, a great wealth of knowledge has been learnt about the surfaces of organelles. In this technique cells are frozen quickly in liquid nitrogen (−196 °C). If the cells are now exposed to a razor blade they tend to fracture along lines of weakness. The fracture plane *often follows the internal membrane system*. An analogous situation occurs when a piece of wood is trapped in the frozen surface of a pond. If the ice is cracked then the fracture plane will invariably pass along the wood – ice interface. One can imagine the surface of the nucleus offering a similar path to the advancing fracture plane. A replica of the surface is then prepared after shadowing with a platinum and carbon mixture. The application of carbon directly from above strengthens the structure and allows it to be removed for observation under the EM. This technique was used in the preparation of the cell membrane seen in Figure 4.8. The so-called intramembrane proteins can be clearly seen.

1.4.5   HIGH VOLTAGE ELECTRON MICROSCOPY

In the conventional transmission EM a potential of around 100 kV acts as an accelerating voltage. In the early 1960s the so-called high voltage EM was developed for scientific work. This form of EM uses a potential of approximately 3×10^3 kV. The electrons produced have sufficient kinetic energy to *penetrate thick specimens or whole cells*. There is no need for ultra-thin sections, and in addition the instrument provides high degree of resolution and a considerable depth of field. Use of this instrument has enabled such features as the cytoskeleton and mitosis to be examined in detail.

1.4.6 SCANNING ELECTRON MICROSCOPY

In recent years the development of the scanning EM has provided biologists with another extremely powerful means of examining surfaces. A relatively low-energy beam of electrons is allowed to strike the surface of the specimen. The electrons are scattered by the specimen and they also cause secondary electrons to be emitted from the specimen. These secondary electrons are gathered by a **Faraday cage**. This instrument contains a **scintillant** which produces photons as a result of collisions with electrons. A system of coils within the column moves the primary beam across the specimen, producing a series of images created by the secondary electrons. The scanning primary beam is synchronised with the scanning of the cathode ray viewing screen. In this way a detailed picture is created. The fibroblasts in Figures 7.13, 7.14 and 7.15b were all photographed using this technique.

1.5   Cell and tissue separation techniques

The eukaryotic cell is a highly ordered structure. Membrane systems of various types essentially create a spatial separation of molecules, allowing a vast number of biochemical reactions within the cell to occur in a controlled manner. Cells exhibit a varying degree of differentiation and this itself creates another order of complexity. Prokaryotic cells, while not approaching this level of organisation, do possess a significant degree of order.

The investigation of cellular processes often involves the isolation of a specific organelle or molecule from this complex and dynamic cellular system. While the cell is intact and living *organelles and metabolites are maintained in their natural state*. The disruption of cells, which must inevitably occur if components are to be isolated, may lead to the *alteration or destruction* of many of these subcellular components.

1.5.1 HOMOGENISATION OF CELLS AND TISSUES

There are many methods routinely in use to homogenise cells and tissues. Just as there is really no typical cell, so no standard method of cell disruption can be said to exist. Cells are most commonly disrupted by some mechanical means. This may be in a glass homogeniser or Waring blender or by forcing cells through a small aperture. Exposure of cells to an osmotic stress or high-frequency sound can also cause disruption.

Regardless of the method used, the destruction of cellular components must be kept at a minimum. Homogenisation is routinely carried out at 4 °C in an effort *to reduce the enzymatic breakdown of cell components*. An osmotic buffer is normally included in the isolation medium if intact organelles are required. Sucrose often fulfils this role, although other chemicals such as sorbitol, ficol or manitol can be used instead. The pH of the isolation medium is also maintained at a specified level. The actual pH depends upon the nature of the final product. Isolation of the enzyme acid phosphatase is normally carried out at pH 5.0 and ribosome isolation medium is usually buffered at approximately pH 7.5.

1.5.2 CENTRIFUGATION

Once the cells or tissues have been disrupted the isolation of purified fractions is most commonly carried out by *centrifugation*. In a centrifuge, particles sediment at different rates when an accelerating force is applied. The rate depends upon the size, density and shape of the particles. A typical cell fractionation scheme is shown in Figure 1.9.

As the particle size decreases, the centrifugation speed must clearly increase to effect the sedimentation of the organelle. Intact nuclei can be separated in a bench centrifuge of the type often found in school laboratories. The isolation of microsomes and ribosomes requires an accelerating force of around 500 000g; such forces are generated in the ultra-centrifuge. The development of this particular form of centrifuge was possible due to the work of the Swedish scientist Theodor Svedberg. The sedimentation coefficients of biological molecules are expressed in Svedberg units (S). A range of sedimentation values is shown in Table 1.1.

Table 1.1 Sedimentation coefficients.

Organelle/molecule	Sedimentation coefficient (S)	Mass (k dal)
bacterial ribosomal RNA	23	1.2×10^6
	16	0.55×10^6
	5	3.6×10^4
transfer RNA	4	2.5×10^4
bacterial ribosome	70	2.7×10^6
eukaryotic ribosome	80	4×10^6

The isolation of a purified mitochondrial fraction in the scheme shown in Figure 1.9 involves sucrose density-gradient centrifugation. The development of this technique was a significant advance in cell fractionation, cytology and molecular biology in general. Two types of gradient can in fact be formed. A continuous gradient is one in which a mixing device creates a smooth gradient, with the bottom being the most concentrated and the top being the lightest. In a discontinuous gradient, a series of discrete density bands are formed by carefully layering one on top of the other.

In the simplest form of density-gradient work the extract is layered on top of a gradient in which the density is less than the particles to be isolated. The application of an accelerating force causes the particles to *migrate in discrete bands according to their density*. The density gradient serves to prevent the spreading of the bands, thus allowing relatively pure fractions to be isolated.

This technique can be modified by centrifuging particles in a gradient which includes the density value of the particle. This is often termed **equilibrium-density** or **isopynic centrifugation**. At equilibrium the particles will be located at a point which is around their own **buoyant density**. In their classic experiment Meselson and Stahl (see §5.3) used a caesium chloride gradient to determine the buoyant density of various DNA strands.

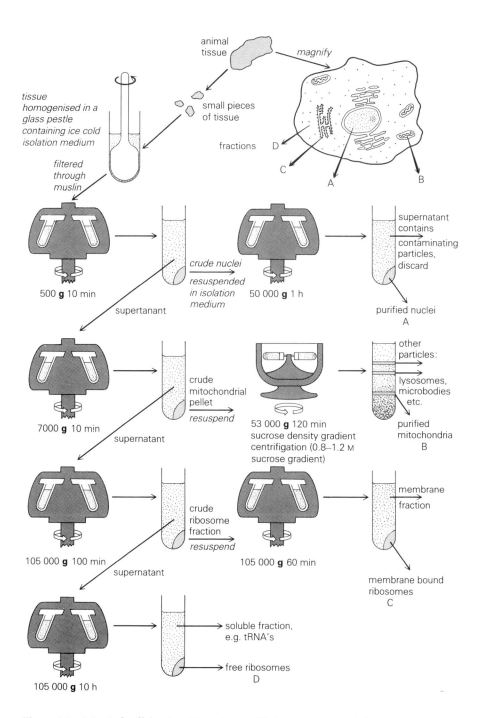

Figure 1.9 A typical cell fractionation scheme. All the steps are carried out at 0–4 °C.

1.6 Chromatography

As we have seen, it is often important to prepare isolated extracts of various cellular organelles in a relatively uncontaminated state. The same is true for the molecular components of the cell. This is achieved most commonly by some form of **chromatography**. There are many different types of chromatography and for convenience four different forms will be considered: **partition**, **adsorption**, **ion-exchange** and **affinity** chromatography.

1.6.1 PARTITION CHROMATOGRAPHY

This technique depends upon the relative solubility of substances in two or more solvents. Such separations can be performed on a variety of chromatography media. One of the most widely used methods is paper chromatography, and consideration of this technique will serve to illustrate the general principles of partition separation. Water molecules are held within the cellulose matrix and act as the **stationary phase** in the partition process. Other substances can provide similar environments in which water molecules can reside, e.g. silica gel, kieselguhr and Celite. In each case the porous medium acts solely as a support for water. The second solvent system passes through or over the chromatography medium and is known as the **mobile phase**. Separation is achieved within the system if the molecules exhibit *different solubilities in the stationary and mobile phases*. In true partition chromatography the speed of movement of any one molecule will be determined solely by its relative solubility in the two solvents (Fig. 1.10).

One can envisage two extreme forms of molecular behaviour. If a particular substance is completely insoluble in the aqueous or stationary phase it obviously follows that all the molecules would be in the mobile phase. The compound would travel through the system at a speed *identical* to that of the *mobile phase*. The opposite situation would exist if a particular compound is completely insoluble in the mobile phase. In this case it will simply *remain at the origin*.

In a given chromatography system the behaviour of a particular molecule in a specified solvent system is quite characteristic. In paper or thin layer chromatography the movement of molecules is expressed in terms of their *position relative to the solvent front*. This value is known as the *relative front*, R_F, and is calculated as follows:

$$R_F = \frac{\text{distance moved by the solute}}{\text{distance moved by the solvent front}}$$

In the case of the two extremes cited earlier, the molecule insoluble in the stationary phase would exhibit an R_F value of 1. The other molecule would be expected to remain on the origin and hence exhibit an R_F value of 0. In any experimental system these two extremes would hopefully be avoided and an intermediate R_F value would be obtained (Fig. 1.10e).

In many biological investigations the effective separation of molecules is seldom achieved after running a chromatogram in a single direction. Two-dimensional chromatography is usually performed. This is carried out using a two solvent system, the chromatogram being run at right angles to the original in the second solvent (Fig. 1.11).

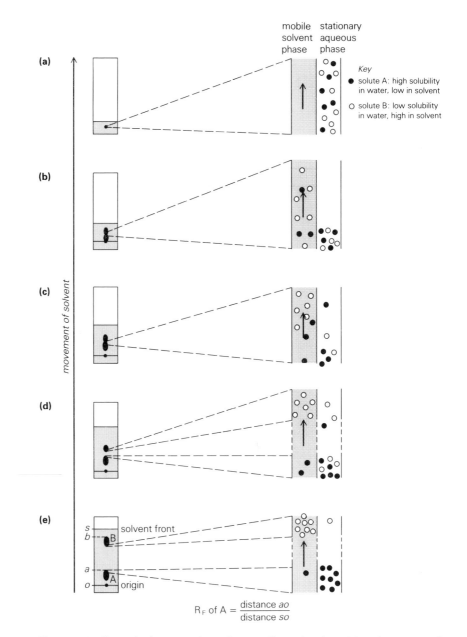

$$R_F \text{ of A} = \frac{\text{distance } ao}{\text{distance } so}$$

Figure 1.10 Stages in the separation using one-dimensional partition chromatography.

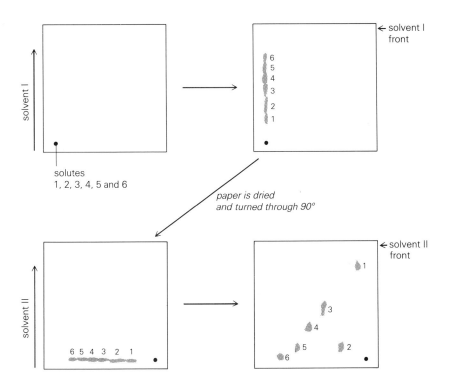

Figure 1.11   Two-dimensional chromatography.

1.6.2   COLUMN CHROMATOGRAPHY

The techniques of paper and thin layer chromatography are excellent analytical tools. However, one of the chief drawbacks lies in their limited loading capacity and the relative difficulty of recovering the separated molecules. The spots can be eluted from the chromatographic medium but this is quite a tedious process. The use of a column chromatographic technique solves both of these problems. This form of chromatography can be modified in a number of ways. In one of the simplest forms it can act purely as a partition system. Here the column is often composed of cellulose, although such compounds as silica gel, Celite or kieselguhr can be used. The mixture to be separated is loaded on the column and the chromatography solvent is passed continuously through the column (Fig. 1.12). The results shown in Figure 1.12c are typical of column chromatography. The use of a phosphocellulose column allows the separation of the RNA polymerase core enzyme from the lower molecular weight sigma factor (see §6.3).

There are many other ways in which separation can be achieved on a column; for example, a molecule sieve can be created. The column is packed with beads of a specific size. These beads, known commercially as Sephadex, act as a sieve through which molecules move according to their molecule weight. Column chromatography of this sort is very commonly used and enables the rapid separation of molecules in quite complex mixtures. The chromatography techniques discussed in the next two sections can of course be carried out on columns.

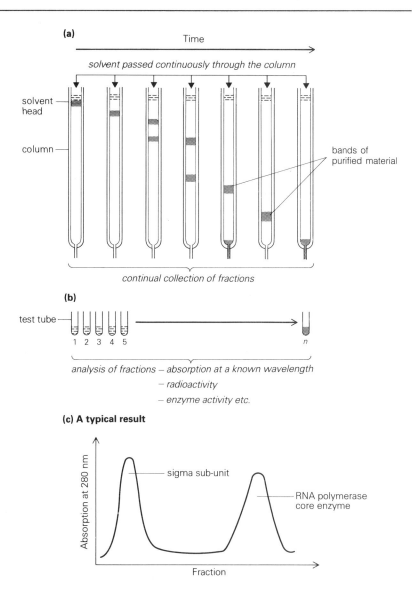

Figure 1.12 Column chromatography. (a) Continuous elution of the column by solvent. This solvent may be changed if ion-exchange chromatography is being performed. (b) Collection of fractions. (c) Separation of RNA polymerase using a phosphocellulose column.

1.6.3 ADSORPTION CHROMATOGRAPHY

Adsorption of molecules onto a solid surface is a widely used technique. Separations of this sort can be carried out on thin layer plates, especially modified paper, or more commonly on columns. In the simplest form the molecules are retarded in their passage through the column by interactions with the medium. These involve weak **van de Waals' forces** or **hydrogen bonding**. If the column is packed with a special resin which contains ionic groups capable of exchange then **ion-exchange chromatography** becomes possible. Such resins consist of small beads which are cross-linked, producing a sponge-like network. Within this matrix, groups may be capable of exchanging cations or anions depending upon the choice of the resin.

A **cation-exchanging resin** contains ionic groups which are capable of exchanging with cations present in the eluting medium. If a solution of molecules is loaded positively molecules such as proteins bind onto the surface of the resin, displacing hydrogen ions. The charged molecules may be slowed in their passage through the column, thus effecting separation, or may remain firmly bound to the resin while contaminating molecules that pass freely through the system. The bound molecules can *be subsequently eluted by changing the pH of the solvent or by increasing the ionic concentration of the column.* **Anion-exchange resins** work in a similar manner.

1.6.4   AFFINITY CHROMATOGRAPHY

The principle of specific binding of molecules to a solid chromatography surface has been extended in the technique of affinity column chromatography. This is perhaps not an unexpected development since biological systems exhibit interactions based on specific molecular interactions. These include antibody-antigen, enzyme-substrate and DNA-RNA interactions. All three of these binding phenomena have been exploited in the separation of specific molecules. The principle can be illustrated by considering the isolation of a specific enzyme. The substrate molecule is linked to a chromatography medium and the mixture of molecules containing the enzyme passed through the column. Under the appropriate conditions the enzyme will bind to the substrate forming an **enzyme-substrate complex**. The elution of the purified enzyme can be achieved by altering the pH of the column, thus reducing the stability of the complex. It is easy to see that this technique can be extended to antibody-antigen and also to nucleic acids. In the latter case, highly specific hydrogen bonding will occur between complementary nucleotide base sequences.

1.7   Electrophoresis

Many biological molecules contain charged groups and this forms the basis of ion-exchange chromatography. It is also the basic principle which underlies **electrophoresis**. If a mixture of charged molecules, e.g. proteins, is placed between two electrodes the molecules will migrate to the oppositely charged electrode. The substances to be analysed are placed in a discrete zone at a suitable distance from each electrode. Electrophoresis is carried out on a stabilising medium and is only rarely performed in free solution. The supporting medium for electrophoresis varies enormously; some are relatively inert, e.g. paper and agar gel. Here separation depends almost totally upon the net effect of charge and mass. Molecules will migrate according to their charge:mass ratio. Use of other media such as the synthetic polymer acrylamide allows a porous electrophoretic medium to be established. The pore size is comparable with the molecules being separated. Increases in gel concentrations produce a decrease in pore size, thus excluding from the gel molecules of a certain size.

1.8   Radioactive techniques

The idea that advances in biological sciences depend upon the developments in the physical sciences is one that we have met previously. In the case of radio-isotopes there is no single advance that one can quote in support of such an idea. It is rather the steady development of techniques and expertise which have aided the investigation of physiological, cellular and biochemical processes. Isotopic labelling has also been used to extend other techniques discussed elsewhere in this chapter, for instance chromatography and microscopy. Indeed the usefulness of these techniques *would be seriously reduced without the use of radio-isotopes*, and the elucidation of many biochemical pathways and subcellular location of important molecules would have proved impossible.

The atomic structure of an atom is extremely complex and not yet completely understood. A simplified representation of a carbon atom is shown in Figure 1.13a. The nucleus contains virtually the total mass of the atom and consists mainly of neutrons and protons. The collective term **nucleon** is used to describe the neutron and proton content of the nucleus. The proton is the only charged particle in the nucleus, and it is the number of protons (Z) which determines the *atomic number* of the element. Electrons move around the central nucleus in orbits determined by their energy levels. These orbits are arranged in shells, and each shell can hold a specific number of electrons. The shells fill sequentially along and down the periodic table. Electrons are negatively charged, and since the *number of electrons and protons are equal in an atom, the whole structure is electrically neutral*. When energy is supplied to an atom, electrons may become excited and move to higher energy levels. On returning to their ground state, energy is released.

An atomic species with a specific nucleon content is termed a **nuclide**. Carbon with six protons and six neutrons is **nuclide carbon 12** whose symbol is $_{6}^{12}$ C, although the subscript 6 is usually omitted. In a population of atoms of a particular element, the vast majority of atomic nuclei will contain equal numbers of protons and neutrons; *however, a small amount will differ*. Thus for any element there exists a family of nuclides in which the atomic number (Z) remains the same, but the mass number differs. These nuclides are termed **isotopes**; examples of biologically useful isotopes are shown in Table 1.2.

Isotopes can either be stable or radioactive. The phenomenon of radioactive decay can be attributed to a variety of interactions within the nucleus of a nuclide. The most important of these in biologically useful nuclides is due to the excess of neutrons. This can be illustrated by considering the isotope

Table 1.2   Biologically useful isotopes.

Element	Radio-isotope	Half-life
hydrogen	^{3}H (tritium)	12.26 a
carbon	^{14}C	5730 a
sodium	^{22}Na	2.6 a
	^{24}Na	15.0 h
phosphorus	^{32}P	14.3 h
sulphur	^{35}S	87.4 d
chlorine	^{36}Cl	3×10^{5} a
potassium	^{42}K	12.4 h

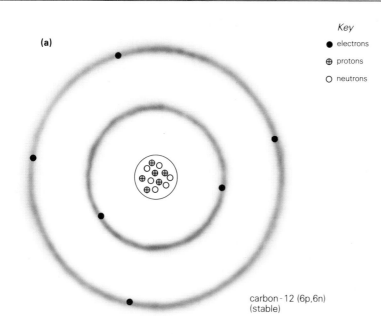

Figure 1.13 (a) Atomic configuration of carbon 12. (b) Decay of carbon 14 into nitrogen 14 with the release of a particle and antineutrino.

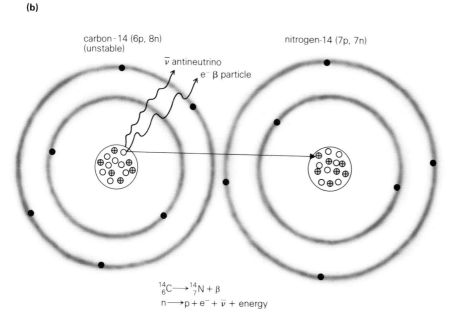

carbon 14. In this case a neutron is converted into a proton by the ejection of a β⁻ **particle** which is an electron with kinetic energy. An **antineutrino** ($\bar{v}$) is also ejected:

$$n \rightarrow p + e^- + \bar{v} + \text{energy}$$

As a result of this process the nucleus has gained a proton and hence moves one place to the right in the periodic table. The majority of biological nuclides are β^- emitters.

The rate at which radioactive isotopes decay varies enormously. It is usual to express this rate of decay in terms of the **half-life** of the isotope. One of the principal factors governing the usefulness of an isotope is the length of its half-life. If this is excessively long as in the case of ^{36}Cl or very short (2 min) as in the case of ^{15}O then the isotope is probably unsuitable for routine investigations.

The detection and counting of radioactive nuclides can now be reliably and efficiently performed in most biological situations. The selection of technique depends upon a variety of factors, e.g. type of particle emitter used, dual or single labelling, immediate source of the radioactive isotope and the experimental procedure employed. An extensive discussion of all the techniques currently used is beyond the scope of this chapter. In order to simplify the problem we might consider that the isotope is located in one of three different situations:

(a) on a solid chromatography medium, thin layer plate or paper chromatogram;
(b) in a liquid medium, e.g. enzyme assay;
(c) in a whole tissue or at a subcellular location.

The detection and counting of a radioactive isotope will clearly be very different for each of these media and the problems posed by the different situations will give the student some insight into the technical limitations of isotope work.

1.8.1   COUNTING ON A SOLID MEDIUM

In many biological experiments, labelled molecules need to be separated by means of chromatography techniques. These techniques are covered elsewhere (see §1.6) and here we are only concerned with the detection of any radioactive molecules after separation. One consequence of radioactive decay is that the emitted particles cause ionisation of other molecules. The most commonly encountered detection device is the **Geiger-Müller** (GM) tube (Fig. 1.14). Such an instrument usually contains an inert gas (helium or argon) which can be ionised by radioactive emissions. The ions which result from these collisions are accelerated by the voltage applied between the cathode and the anode. These ions may acquire sufficient energy to cause *further ionisation events creating a cloud of secondary ions*. These ions can in turn create further ions as they collide with gas molecules. This process is known as **gas amplification** and it considerably increases the sensitivity of the instrument. However, it does also mean that the GM tube is incapable of discriminating between different types and energies of emission. The creation of secondary, tertiary and subsequent ionisation events inevitably *means that the final current pulse is independent of the number of primary ions*.

Another problem arises since the positive ions which are created are relativly slow and they tend to build up inside the chamber. Until they are cleared out of the system any further ionisation events are undetectable. This period is known as '**dead time**' and can be as long as 200 μs. The result of this is a *significant loss in counting efficiency*. Further reduction in counting efficiency results from back-scatter, self-absorption within the source,

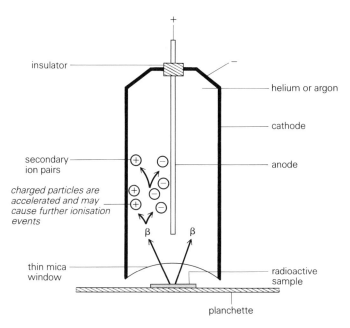

insulator

helium or argon

cathode

secondary ion pairs

anode

charged particles are accelerated and may cause further ionisation events

β   β

thin mica window

radioactive sample

planchette

Figure 1.14 Cross section of a typical thin end window Gieger–Müller tube. Gas amplification is indicated by the creation of secondary ion pairs.

absorption by the counting window and absorption in the air gap between the source and the detector. Although these factors amount to significant decreases in counting efficiencies, the GM tube remains a powerful tool when analysing thin layer and paper chromatograms.

A wide range of specialist instruments exists to scan such chromatograms. Many of these are based on the windowless GM tube. In such apparatus the gas is only partially enclosed, a thin slit remaining open in the detection area. Gas is allowed to flow into the detection chamber and out of the slit. Counting efficiencies with such instruments tends to be between 10-15 per cent for ^{14}C but below 2 per cent for ^{3}H.

The usefulness of thin layer chromatography and paper chromatography can be increased by **autoradiography** (see §1.8.4) or by elution of the labelled molecules from the solid supporting medium. The radioactive components can then be counted in a liquid phase by the technique of liquid scintillation counting.

1.8.2 LIQUID SCINTILLATION COUNTING

In this method the radioactive sample is dissolved or suspended in a liquid scintillant. The scintillant contains a compound (phosphor) which is capable of emitting light photons following the absorption of ionising radiation. The photons created strike a photocathode, causing the emission of electrons (Fig. 1.15a). These electrons are attracted towards a dynode, liberating **secondary electrons**. The electrons in turn collide with secondary dynodes creating additional electrons. This process is repeated many times leading to a final amplification of 10^{6}-10^{7}. The resultant stream of electrons is further amplified before detection.

The liquid scintillant contains three essential components. The primary solvent is an aromatic solvent, usually toluene or xylene. The primary solute or phosphor is most commonly either PPO or butyl-PBD (Fig. 1.15b). The

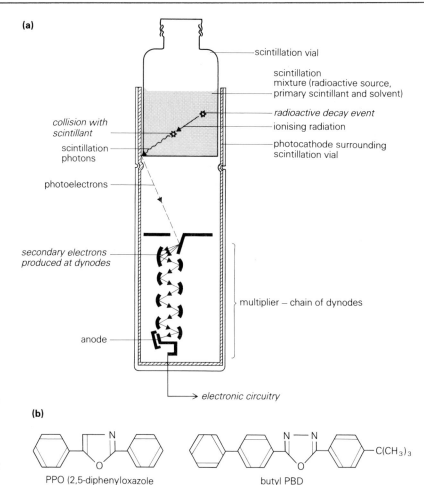

(a)

scintillation vial

scintillation mixture (radioactive source, primary scintillant and solvent)

radioactive decay event
ionising radiation

collision with scintillant

scintillation photons

photocathode surrounding scintillation vial

photoelectrons

secondary electrons produced at dynodes

multiplier – chain of dynodes

anode

electronic circuitry

(b)

PPO (2,5-diphenyloxazole

butyl PBD

Figure 1.15 (a) Scintillation detector and photomultiplier tube. The photocathode is cut away to show the position of the scintillation vial. The stream of electrons is further amplified and used to operate counting equipment. (b) Chemical structure of primary solutes PPO and butyl-PBD.

ionisation events caused by the radioactive emission excites electrons in the phosphor and photons are emitted as the electrons return to their ground state. These photons then initiate the events described above. Unfortunately, most radioactive biological samples have a low solubility in toluene or xylene and a **secondary solvent** or **solubiliser** needs to be added to the scintillant. Ethanol is often incorporated into the mixture in a ratio of 7 : 3 toluene : ethanol. Such a mixture allows 0.2 cm³ to be incorporated into the scintillant. Greater volumes of the aqueous phase can be added by use of an emulsion based on one of the commercially available compounds such as Triton X-100. The major, yet clearly unavoidable, consequence of a secondary solvent is the *lowering of the counting efficiency*. This effect is known as '**quenching**' and great care must be taken to minimise this phenomenon. There are many factors which may lead to quenching. The addition of a coloured component or particulate matter will quench owing to absorption of the photons by the sample itself. Modern scintillation counters are, however, capable of making sophisticated quench corrections.

Liquid scintillation allows counting efficiencies of approximately 70 per cent for ^{14}C and 20 per cent for ^{3}H. In addition, detection of individual isotopes within dual or triple labelled samples is possible.

1.8.3 TISSUE SOLUBILISATION AND COMBUSTION

Frequently the radioactive compound is within a medium which prevents the use of any of the techniques discussed. This may occur when the isotope is located in an animal or plant tissue or in a polyacrylamide gel (see §1.7). Here the supporting medium may need to be solubilised by the use of a commercially available tissue solubiliser, e.g. Digestin or Protosol. The use of such compounds requires care *as they may cause severe quenching problems or may even affect the stability of the scintillant.* An alternative approach is to oxidise the extract and collect the radioactive combustion products, ^3H_2O and $^{14}CO_2$. These can then be counted using conventional scintillation techniques.

1.8.4 AUTORADIOGRAPHY

In the previous sections, methods of counting radioactivity have been covered. It is frequently important to determine the location of an isotope within an organism or at a subcellular site. Radioactive emissions will create a track when they pass through a photographic emulsion. The development of the emulsion reveals the location of this track. This forms the basis of **autoradiography**.

The technique of autoradiography can take several forms. The simplest method is known as **apposition**. The specimen (whole organism, tissue section or chromatogram) is dried and placed against an X-ray plate. Development of the plate after a suitable period of time results in areas of exposure corresponding to the radioactivity within the specimen. This technique is used to detect the location of radioactive assimilates in plants.

When autoradiography of microscopic sections is to be undertaken, a photographic emulsion is overlaid on the specimen. As the emulsion dries it adheres closely to the surface of the tissue. The close proximity of the emulsion to the cell surfaces makes the detection of the weak β emitter tritium possible. The particles emitted from this isotope only travel 1 or 2 μm and hence are useful in **electron autoradiography**. In such high-resolution studies *the use of more powerful emitters would tend to produce artefacts* (Fig. 1.16a). The chromosomes in Figure 1.16b were subjected to autoradiography after hybridisation with tritium-labelled 5 S ribosomal RNA. Complementary sequences on the chromosome can be recognised as areas of exposure in the emulsion.

1.8.5 PULSE-LABELLING TECHNIQUES

In many cases biological molecules undergo considerable modifications after synthesis. The most well documented case of this occurs in the production of nucleic acids. In the case of ribosomal RNA (see §6.2.1) a large transcription product is produced and subsequently cleaved to produce mature molecules. The precise nature of this process has been elucidated by **pulse-labelling** techniques.

In this experimental procedure actively growing cells are exposed to a radioactive precursor for a short period. The labelled precursor is then removed and replaced by 'cold' (unlabelled) precursor molecules. The unlabelled precursors are incorporated into the newly synthesised molecules and have the effect of 'chasing' the previously synthesised molecules

(a)

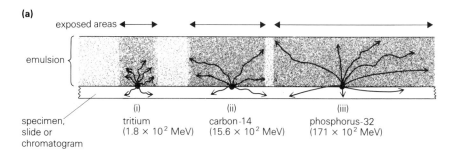

(b)

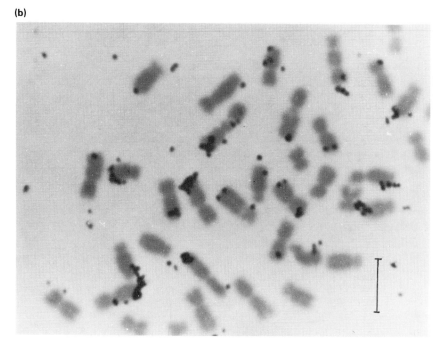

Figure 1.16 (a) Particle emission from various sources. The likely area of emulsion exposure is indicated. The low-energy particles produced by tritium make it ideally suited to high-resolution autoradiography. (b) Autoradiograph of chromosomes from *Xenopus laevis* kidney cells maintained in culture. The chromosomes have been hybridised with tritium-labelled 5 S RNA (scale bar 5 μm).

containing the radioactive precursor through any maturation process (Fig. 1.17). If the molecular species under investigation is sampled shortly after the start of the experiment, *only the primary synthetic product will contain radioactivity*. After longer time intervals, the original radioactive molecules will have been replaced with non-radioactive molecules. The radioactive molecules will have changed and perhaps even have moved away from their site of synthesis. In this way the flow of radioactivity through a maturation process can be followed, together with any movement of molecular species within the cell.

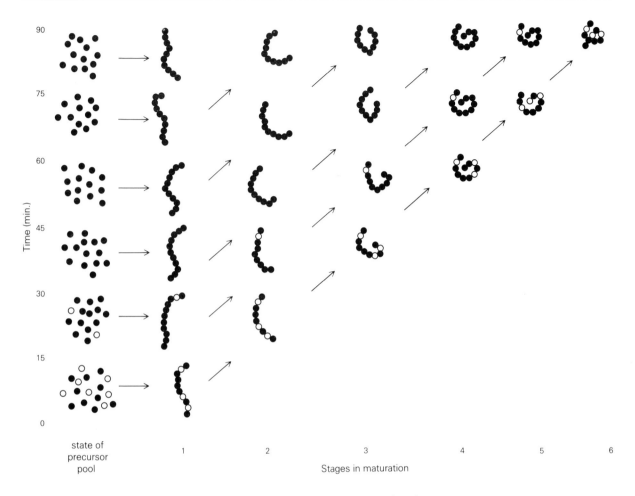

Figure 1.17 Pulse-labelling technique. Radioactive precursors (o) are introduced into the system for a short period. These are then incorporated and chased through any maturation process by non-radioactive precursors (•).

1.9 X-ray diffraction

X-rays have wavelengths of the same order as the distance between atoms. This has enabled X-rays to be used as a tool *to probe the spatial arrangement of atoms in biological molecules.* The position of atoms in a regular crystal lattice acts like a **diffraction grating** and is effective in producing an X-ray spectrum. The same effect can be obtained when a ruled diffraction grating is used with visible light.

In **X-ray crystallography** a beam of electrons with a wavelength of 15 nm is accelerated towards the crystal. Part of the beam is scattered by the atoms in the crystal and these rays can be detected by means of photographic film. The extent and nature of the scattering will be *related to the orientation of the* *atoms on the crystal*. The diffracted beams may recombine and the resultant phase depends upon the same rules which governed visible light in phase contrast microscopy (see §1.3.2). The X-rays that are in phase will reinforce each other and the resultant amplitude will be greater than either of the two individual component waves.

Movement of the crystal in the X-ray beam produces a series of diffraction pictures. Using data obtained from the intensity of the individual spots and the phases of the scattered beams, a complex analysis of the molecule can be performed. This is achieved by a mathematical analysis known as the **Fourier series**. This synthesis gives the density of the electrons within the crystal. An electron density map can be constructed by tracing the two-dimensional electron maps on to transparent sheets, with each sheet representing a different plane within the crystal structure. The use of modern computing techniques has considerably speeded up this process.

The first proteins to be successfully analysed by the technique were myoglobin and haemoglobin. The analysis of myoglobin by John Kendrew in 1957 produced a diffraction pattern at a 60 nm resolution. Five years later Kendrew had successfully produced an analysis of a 14 nm resolution. In the last 20 years the structures of over 50 proteins have been elucidated by this technique. In addition the methods have been extended and refined, enabling the structure of other molecular types to be analysed. Any consideration of X-ray diffraction would be incomplete without a mention of the analysis by Watson and Crick of the DNA helix by this technique. The repetitive nature of the DNA molecule makes it an ideal structure for such analysis. A more detailed discussion of this molecule and its structure can be found in Chapter 5.

Cell Biology

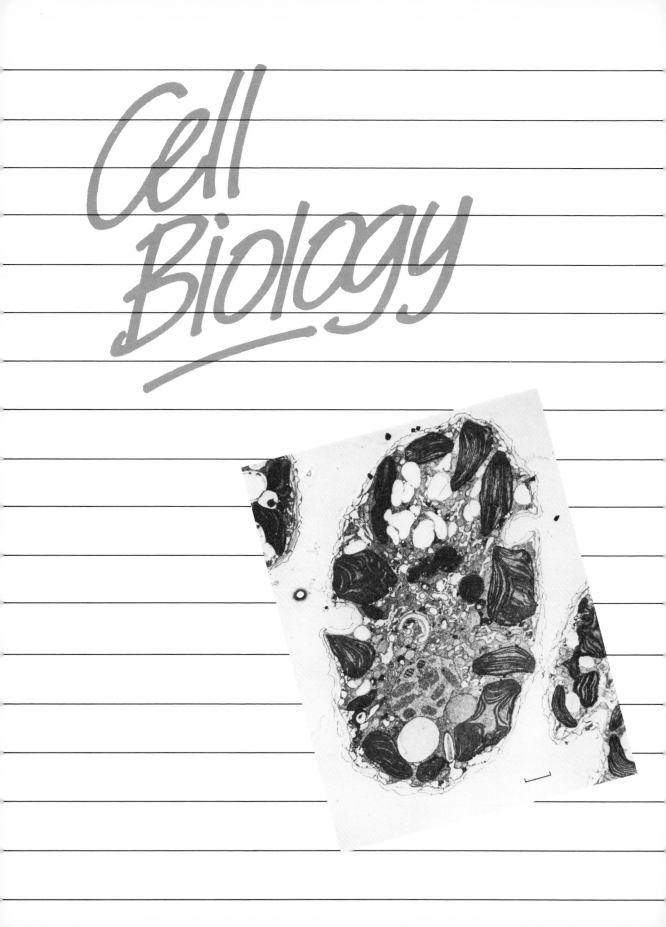

2

Evolution of cells

M. A. Sleigh

MODERN VIEWS IN BIOLOGY

2.1 The essential characteristics of cells

2.1.1 MEMBRANES

All cells are delimited by a membrane composed of a bilayer of lipid molecules, associated with a diversity of proteins. These proteins contribute to such functions as selective permeability, transmembrane transport and structural integrity (see §4.1.2). This membrane provides a selective barrier to the passage of substances into and out of the cell, and thereby controls the concentrations of molecules within the cell; *some concentrations are regulated passively and some by active uptake or extrusion.* This control, which may be supplemented by specific regulatory functions of intracellular organelles, *provides the environment in which internal enzymes work.* The cell cytoplasm therefore consists of a watery solution containing controlled concentrations of various ions and a diversity of organic compounds including monomers and polymers of sugars, amino acids, fatty acids and many intermediates of the metabolic processes of the cell. Materials may be actively secreted outwards through the surface membrane (see §4.2.1), either as substances that remain associated with the cell surface as a cell coat (**glycocalyx**) or cell envelope, wall or capsule, or as substances that the cell exports as waste products or to function as enzymes or antibiotics.

Within the membrane are cell components concerned with the metabolism of the cell and providing both energy and materials needed for growth and reproduction. All organisms make use of energy carriers in the form of nucleoside phosphates (most commonly adenosine triphosphate, ATP), and cells must contain mechanisms for the synthesis of ATP, usually by the phosphorylation of ADP. This is achieved in a variety of ways, depending on the availability of oxygen and the type of organism involved. The energy provided by these carrier molecules is used in a wide range of enzyme-catalysed reactions. These include the synthesis of macromolecules, the transport of molecules or ions through membranes and molecular shape changes that may be coupled to motility. Some groups of enzymes are involved in the catalysis of chain reactions, and tend *to be maintained in a fixed spatial relationship to one another by attachment to a membrane;* many other enzymes, such as those concerned with transport through membranes, are also associated with or embedded in membranes.

2.1.2 DNA, RNA AND PROTEIN SYNTHESIS

The specificity of reactions catalysed within the cell is determined by the enzymes that promote these reactions (see Ch. 3, page 117). In those synthetic reactions where the composition of the molecule that is synthesised is complex, comprising a specific sequence of nucleotides in a nucleic acid molecule or a specific sequence of amino acids in a protein molecule, not only must specific enzymes be present but *there must also be a template molecule that indicates the sequence in which components must be incorporated.* The organism must inherit these template molecules, or the ability to produce them, *in order that the offspring can have the same synthetic capability as the parents.*

This inheritance resides in the sequence of nucleotides of the DNA of chromosomal genes in the cell; this sequence can be copied into further DNA chains to be passed to the progeny of the cell, or can be copied into RNA ('messenger') molecules whose nucleotide sequence can be used to provide instructions for the sequence of amino acids in a protein. *Every cell must therefore contain a set of genes embodying the instructions for synthesis of every enzyme and other protein found in that cell.* It must also contain synthetic machinery capable of replicating the genetic DNA (see §5.3). In addition it must have enzymes capable of **transcribing** a DNA sequence into a messenger RNA sequence and secondly of **translating** the messenger RNA sequence into an amino acid sequence during the synthesis of a protein (see §6.8). *All types of cell synthesise proteins on cytoplasmic ribosomes,* which themselves comprise (ribosomal) RNA and protein. Transcription also requires specific linking (transfer) RNA molecules to match specific amino acids to the coded components of the messenger RNA. The synthesis of these three types of RNA, as well as of replicated DNA chains, all takes place at the chromosomal genes.

2.1.3 EUKARYOTES AND PROKARYOTES

The cells of different types of organism vary greatly in the extent of membrane systems and fibrous systems that exist within the cell membrane (Figs 2.1a & b). Simple bacterial cells like *Mycoplasma* contain neither internal membranous structures nor intracellular protein filaments, whilst the cells of higher plants and higher animals contain both complex membranous organelles and extensive systems of protein filaments that may form a complex cytoskeleton within the cell (see §7.3). Cells can be divided into two groups according to the presence or absence of a membranous envelope around the chromosome(s) that provides a nuclear compartment within the cell. Those organisms whose cells lack a membrane around the genetic material, and in which the single chromosome is merely in a special region of cytoplasm (the **nucleoid**), are referred to as **prokaryotes**; these are the bacteria and their relatives. Cells of other organisms, the **eukaryotes**, normally have several chromosomes enclosed within an envelope of two membranes containing pores, through which communication takes place between the cytoplasm and the interior of the nucleus. There are also differences between prokaryotes and eukaryotes in the chemical organisation of the chromosomes and their replication, as well as in processes involving RNA, which will be mentioned later.

Most eukaryote cells possess **mitochondria** and many possess **chloroplasts**. These are complex membranous organelles enclosed by two membranes, and concerned respectively with the oxidative phosphorylation steps of respiration and with photosynthesis. Prokaryotic cells do not contain such complex organelles; in those prokaryotes that practise *oxidative phosphorylation or photosynthesis, the components responsible are either bound to the surface membrane or are associated with simple membranous sacs.* Those fibrous proteins that form microtubules or microfilamentous components of the cytoskeletons of eukaryote cells *are seldom, if ever, present in prokaryotes,* and the cytoplasmic streaming movements that tend to be associated with the presence of fibrous systems in eukaryote cells *are not seen in prokaryotes.*

(a)

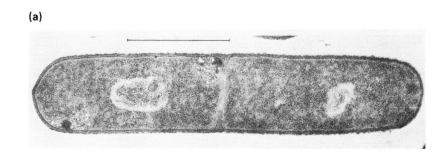

(b)

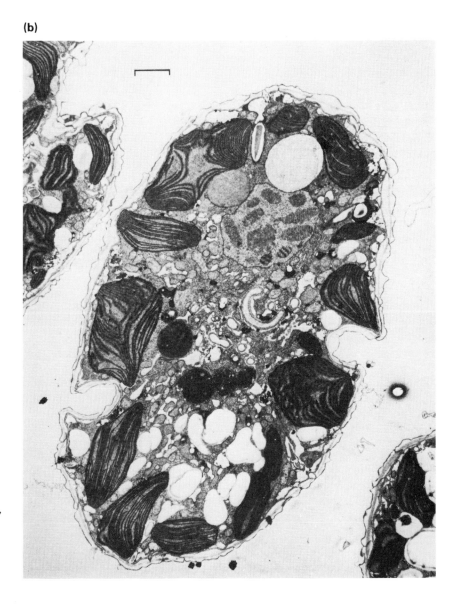

Figure 2.1 (a) LS *Bacillus subtilis* dividing. The cross walls are evident as are nucleoids and vesicular mesosomes (scale bar 1 μm). (b) Eukaryotic cell organisation. Dividing cell of a dino flagellate, showing nucleus, plasted mitochondria and other membranous inclusions within a complex pellicle (scale bar 2 μm).

2.2 Origin and evolution of early cells

It is believed from geological evidence that the Earth is a little more than 4500 million years old. The earliest well formed fossils of higher animals and plants are dated at about 600 million years ago, but fossils of lower organisms have been described from rocks as old as 3000 to 3500 million years (Fig. 2.2). The *first organisms are thought to have originated about 3500 million years ago in conditions rather different from those that exist now,* and it is interesting to consider how living organisms developed and what those first living things were like.

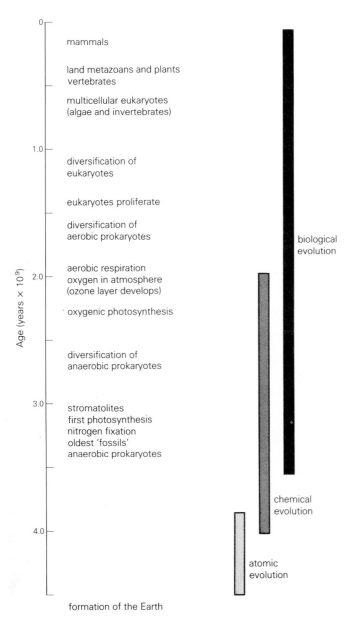

Figure 2.2 The chronological sequence of cellular evolution, based on evidence from fossils and geochemistry

The transition from complex chemical systems existing before the origin of life to the first living cells must have been gradual. However, it seems essential that the first proper organisms *must have possessed a selectively permeable membrane, a mechanism for replication of template molecules and specific synthesis of enzyme molecules, and a source of energy to drive synthetic reactions.* It would not have been essential for any of these components to have been quite as they are in present-day cells.

2.2.1 CHEMICAL EVOLUTION

The atmosphere of the early Earth is believed to have consisted largely of hydrogen and helium with small amounts of other gases like nitrogen, but these were supplemented with various molecules like H_2O, CH_4, CO_2, CO, NH_3 and H_2S outgassed from the Earth's crust. The *amount of free oxygen is assumed to have been negligible, and the atmosphere was a reducing one.* When gas mixtures containing H_2, CH_4, NH_3, H_2O and H_2S were experimentally placed in a closed vessel in the absence of oxygen and exposed to such intense energy sources as spark discharges or ultra-violet light for periods of several days, a mixture of organic molecules formed in the vessels. These molecules included several amino acids, aldehydes and a variety of other organic acids. *Plausible pathways lead from such initial compounds to the synthesis of a full range of sugars, amino acids, lipids and components of nucleic acids.* In the presence of free oxygen such molecules would gradually oxidise, but in a reducing environment they might have persisted. Under certain conditions, such as when locally concentrated by evaporation or adsorption, polymerisation may have occurred forming peptides, nucleic acids, polysaccharides, etc. By these routes, then, it is believed that a wide range of organic molecules came to be synthesised **abiologically** over very long periods of time on the early Earth, during a period of chemical evolution (Fig. 2.2).

2.2.2 BIOLOGICAL EVOLUTION

Mixtures of different forms of organic molecules in solution have been shown to form aggregates called **coacervates** that can accrete molecules from the environment and grow as bodies segregated from the main bulk of the solution. *If coacervate-like aggregates included lipid components for the surface, and contained RNA template molecules capable of self-replication and the synthesis of a few simple enzymes, then the basis of life could be present.* Such an ancestral organism (a '**progenote**') could grow by incorporation of more organic molecules from its surroundings and gain energy for its own synthesis of essential RNA and enzymes by breaking down some of the organic molecules it 'ate'. A primitive organism of this type would be a **heterotroph** (an organism that gains energy from the breakdown of organic molecules imported into the organism from outside); it would be dependent upon a continuous supply of appropriate organic molecules in the surrounding medium. It would also be engaged in some synthesis of organic molecules – *the era of biogenic synthesis of organic compounds had begun.*

In time such organisms could have developed more and more complex patterns of synthesis, using the heterotrophic energy to convert simple carbon compounds like CO_2 into organic molecules and to fix atmospheric nitrogen into amino acids and other organic nitrogen compounds. They therefore became more demanding of energy. The *abiological synthesis of organic compounds must have been slow*, and in the presence of a growing number of heterotrophic organisms making increasing demands on the supply of organic compounds, *the utilisation of organic material must sooner or later have exceeded the rate of synthesis*. There must therefore have been great competition to capture more organic food or to exploit alternative sources of energy for synthetic processes. Light provides an abundant source of energy, *and some organic compounds absorb light*. Organisms that could make use of light in some way to drive chemical reactions could become independent of abiologically synthesised compounds. The use of light energy to drive cell metabolism by photosynthetic organisms (**autotrophs**) allowed a *substantial acceleration of the biogenic synthesis of organic matter*. This provided in turn more food for heterotrophic organisms, enabling them to become predators, parasites or post-mortem decomposers of the autotrophic forms. Hence an **adaptive radiation** of progenotes and early prokaryotic organisms took place to exploit the diversity of environments and energy sources available to them, and *the adaptive evolution of prokaryotes must be assumed to have continued ever since*.

The evolution of the progenote appears to have been relatively rapid, considering the very significant development in organisation of living organisms that took place in the time that passed. It has been suggested that these rapid changes were possible because the organisation lacked rigid controls and hence diverse biochemical experiments were possible in the different lineages of organisms. The transition from a progenote to a prokaryote might be regarded as being marked by an *increasing complexity of the genetic mechanism, with the adoption of DNA as the primary site of inheritance of information for synthesis of the RNA molecules that function in turn in protein synthesis*. It seems likely that there were a number of lines of early prokaryote, since apparently at least three survive today, having given rise to **archaebacteria**, **eubacteria** and the '**urkaryote**' ancestor of eukaryote cells (see §2.7).

2.2.3 PRODUCTION OF OXYGEN

At first, photosynthesis is likely to have been of a primitive type that did not liberate oxygen. The organisms responsible are believed to have been the living parts of **stromatolites** that have occurred in the fossil record since at least 3000 million years ago (and in present-day forms contain **cyanobacteria**, blue-green algae). Eventually, more advanced forms of photosynthesis in which water is split and oxygen released were developed by prokaryotes, probably cyanobacteria. The release of oxygen by these photosynthetic cyanobacteria first allowed the oxidation of reduced compounds of the Earth's crust (an event dated at a little over 2000 million years ago) which led to the release of oxygen into the atmosphere. *The original reducing atmosphere was thereby destroyed and conditions suitable for abiological synthesis were abolished*. Organisms that could only survive in conditions without oxygen (**obligate anaerobes**) were thereafter confined to such habitats as decomposing muds where little or no free oxygen existed; such habitats continue to be inhabited by anaerobic bacteria to this day.

Other prokaryotes adapted to the presence of oxygen and even exploited it. The liberation of energy by the breakdown of organic molecules in anaerobic respiration yields relatively little energy, in comparison with the aerobic oxidative phosphorylation processes associated with the **Krebs Cycle system** and the **electron-transport chain**. Some prokaryotes developed aerobic respiration using these enzyme systems and became more efficient hetero-trophs, but many other prokaryotes, including photosynthetic ones, continued to use simpler respiratory routes and either remained obligate anaerobes or became tolerant of aerobic conditions.

The presence of oxygen in the atmosphere also led to another change in the environment on Earth. Ultra-violet light is absorbed by various types of molecule; for example nucleic acids and proteins absorb it strongly, *and the energy absorbed causes the disruption of chemical bonds, damaging the molecules* (see §5.3.3). Water also absorbs ultra-violet light to some extent, so the early organisms could live well below the water surface without damage to their vital molecules, *but would be damaged in shallow water or on land*. Oxygen also absorbs ultra-violet energy and is converted to ozone. Some of the oxygen in the early atmosphere was converted to ozone and an ozone layer was formed in the upper atmosphere, where it persists to this day. This ozone layer absorbs a large part of the ultra-violet light energy that approaches the Earth from the Sun, *and this reduction in ultra-violet irradiation made it possible for organisms to live in shallow water or on land without suffering disastrously from the damaging effects of ultra-violet light.*

2.2.4 EUKARYOTES

In these conditions, about 1500 million years ago, the eukaryotes became important, most containing organelles capable of aerobic respiration (the mitochondria) and many with organelles capable of photosynthesis (the chloroplasts). They also had the ability to produce microtubular fibrils that take part in segregation of chromosomes in nuclear division (**mitosis**), and that form the axis of cellular projections with intrinsic motility (**flagella**). At an early stage these organisms developed sexual features involving the fusion of two cells and fusion of their nuclei; this doubled the number of chromosomes, and was normally followed by a special type of nuclear division (**meiosis**) in which the chromosomes separated again into two groups, restoring the original number but in new combinations. The processes of mitosis and meiosis have some variants (see §2.9.1), but follow essentially the same pattern in most, if not all, eukaryotes, *so they are assumed to have evolved early in the development of eukaryotes.*

The eukaryotes diversified in an adaptive radiation of protists. The first of these are assumed to have been heterotrophs gaining organic food by phagocytically engulfing prokaryotes, or by absorbing dissolved organic materials. Some of these protists came to possess chloroplasts of different types and became autotrophs. These protists have led on to red algae, brown algae and several important groups of microscopic protists as well as to the green algae that later gave rise to the green land plants. Certain heterotrophic protists evolved into fungi and multicellular animals as well as surviving as a variety of lines of protozoan organisms. Members of various groups of protist are represented in the fossil record since 1400 million years ago, fossilised animal skeletons appeared about 700 million years ago, and fossils of land plants and land animals about 400 million years ago (Fig. 2.2).

If we review present understanding of the history of the Earth and the history of living organisms upon it, we see that there was *no life during the first 20 per cent of the* present age of the Earth. Following the appearance of the first living organisms, the period when life on Earth was *dominated by* *prokaryotes continued for a further 45 per cent of the Earth's history*. The subsequent rise of importance of *protists during the next 20 per cent of the* age of the earth led into the *dominance of higher animals and plants during* *the last 15 per cent* of Earth history (Fig. 2.2).

2.3 Prokaryotes: archaebacteria and eubacteria

The cells of prokaryotes are usually no more than a micrometre or so in diameter, including their secreted cell walls, and seldom show much interior structural detail in electron micrographs (Fig. 2.1a). Prokaryotes are therefore structurally relatively homogeneous, *but their biochemical activities, and the* *ecological niches for which their metabolic specialisms equip them, are* *extremely diverse*. Attempts to unravel their evolution, or even their possible relationships, cannot depend to any extent on structural features, so that attention has turned to biochemical similarities and differences. Proposals of relationships based upon such trends require collateral support, and among the most promising techniques for studying the evolutionary distance between different organisms are those that examine the similarity of amino acid sequences in proteins or nucleotide sequences in nucleic acids in cases where these molecules are both universally present and **highly conserved** (subject to little variation). The sequence of amino acids in cytochrome c has been used to study the closeness of relationships among eukaryotes, and such studies have more recently been extended to prokaryotes. An even more suitable molecule seems to be the medium-sized RNA component of the ribosome, identified as the 16 S RNA of prokaryote ribosomes and the 18 S RNA of eukaryote ribosomes (16 S and 18 S indicate their sedimentation rates in the ultracentrifuge and relate to molecular size (see §1.5.2 & §6.2.1)). This type of RNA turns out to be a very stable molecule associated with the machinery for protein synthesis, rather than a variable type concerned with the production of a specific type of protein.

Analysis of 16 S ribosomal RNA sequences of prokaryotes indicates that the organisms form two distinct groups. Within each group there are similarities of sequence, *but sequences in members of one group are as different from* *those in members of the other as either of them are from sequences in* *eukaryote 18 S ribosomal RNA*. The large majority of prokaryotes form a group called the eubacteria (true bacteria), while the other, smaller, group are the archaebacteria. Biochemical study of members of these two groups shows several other consistent differences. The membrane lipids of living organisms consist of two hydrocarbon chains attached at one end to a glycerol molecule; in both eubacteria and eukaryotes there are two straight chain fatty acids attached to the glycerol through an ester

link (Fig. 2.3a), but in archaebacteria there are two branched chain hydrocarbons (phytanols) attached to the glycerol through an ether (—O—) link (Fig. 2.3b). The cell walls of eubacteria consistently contain the sugar

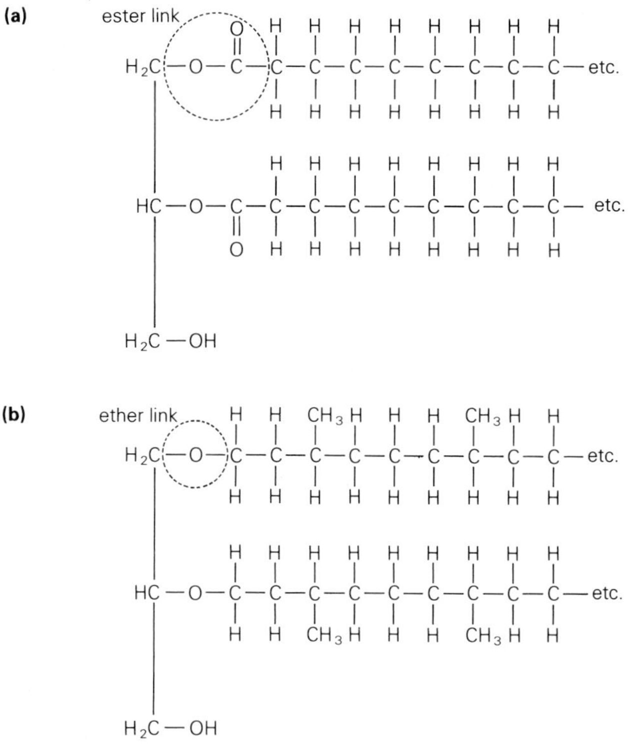

Figure 2.3   Membrane lipids of (a) eubacteria and (b) archaebacteria.

derivative *muramic acid*, while the composition of cell walls of archae-bacteria varies *but never includes muramic acid*. There are also some more detailed differences of transfer RNA composition and ribosomal functioning between eubacteria and archaebacteria, and it is interesting that in a few aspects the archaebacteria are more similar to eukaryotes than they are to eubacteria. This evidence supports the idea mentioned earlier (see §2.2.2) of *an early evolutionary divergence of eubacteria, archaebacteria and the urkaryote*.

2.4   Evolution of energy metabolism among prokaryotes

The driving force for evolution is the need to capture energy more successfully than competing organisms, and the evolution of prokaryotes can be followed through the *exploitation of biochemical processes for more efficient interception and use of energy available in their environment*. Energy is stored within organisms in the chemical bonds of organic molecules, and may be mobilised for use by the oxidation of organic compounds under controlled conditions which allow the transfer of energy to the phosphate bonds of such energy carrier molecules as ATP or guanosine triphosphate (GTP). Most of these oxidation reactions involve the transfer of electrons from an electron donor to an electron acceptor by passage of the electrons through an electron-transport chain of metalloproteins and other 'coenzymes'. Each component in the chain is first reduced and then oxidised by accepting and then donating electrons in a series of coupled oxidation-reduction reactions; it will be helpful to consider some characteristics of these reactions.

When an electron donor (reducing agent) is oxidised by the loss of electrons, these electrons are gained by an electron acceptor (oxidising agent) which is reduced during the coupled reaction (Fig. 2.4). The oxidation of a reducing agent is a reversible reaction and the reduced and oxidised forms comprise a **redox pair**; similarly the reduced and oxidised forms of the oxidising agent comprise a redox pair. In a coupled reaction, the redox pair that has the *stronger tendency to donate electrons (has a lower affinity for electrons) acts as the reducing agent*. The tendency of a redox pair to donate electrons can be related to the *redox potential* of the pair which ranges from a high negative value in strong electron donors to a high positive value in strong electron acceptors (Table 2.1). A redox pair standing higher on this scale can be oxidised by a redox pair standing lower on the scale; *the greater the difference between the redox potentials of the two participating pairs (i.e. the further apart on the scale), the greater the energy released during oxidation.* Synthetic (reductive) reactions generally involve participation of components with strong reducing power, such as NADH or NADPH, which lie quite high on the scale. The oxidation of substrates by donation of electrons also normally involves the removal of protons (H^+ ions) from the compound that is being oxidised (dehydrogenated), and protons are also normally gained by the ultimate electron acceptor (Fig. 2.5); protons may or may not be gained and lost by intermediates in the chain, according to the nature of the compound, for there is a ready supply of protons in the cytoplasm. When electrons are passed from an electron donor to an electron acceptor down a chain of metalloproteins and coenzymes (acting as coupled redox pairs) in cellular electron-transport chains, the component steps regulate energy release and allow *conservation of energy by coupling energy release in some of the steps directly or indirectly to the phosphorylation of ADP* (Fig. 2.5).

Several groups of organic compounds were important for evolutionary progress in the energy metabolism of prokaryotes. Probably the first

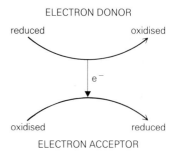

ELECTRON DONOR

reduced   oxidised

e^-

oxidised   reduced

ELECTRON ACCEPTOR

Figure 2.4 The basic form of an oxidation–reduction reaction involving the transfer of an electron e^-.

Table 2.1 Standard redox potentials (E_0') at pH 7 and 25 °C of some redox pairs important in cell biochemistry. Various cytochromes range between −0.10 and +0.36.

Redox pairs	E_0' (V)
acetate + CO_2/pyruvate	−0.70
formate/CO_2	−0.42
H^+/H_2	−0.42
ferredoxin ox./red.	−0.41
NAD^+/NADH; $NADP^+$/NADPH	−0.32
sulphur/H_2S	−0.22
FAD^+/$FADH_2$	−0.22
fumarate/succinate	+0.03
ubiquinone ox./red.	+0.10
nitrate/nitrite	+0.42
oxygen/H_2O	+0.81

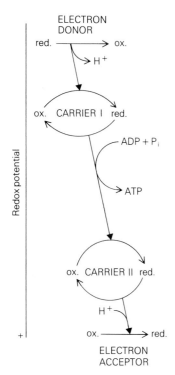

Figure 2.5 A simple electron-transport chain linking the oxidation of an electron donor (at a negative redox potential) with the reduction of an electron acceptor (at a positive redox potential). Intermediate carriers (metalloproteins or other co-enzymes) are first reduced and then oxidised as electrons pass along the path indicated by arrows with solid heads. One step is coupled to ATP synthesis, and protons are lost by the electron donor and gained by the electron acceptor.

metalloprotein involved in cell metabolism was ferredoxin (a simple iron-sulphur protein). The related porphyrins, which are thought to have been developed from ferredoxin by **gene duplication** (§3.2.2), include iron porphyrins in cytochromes and in the enzyme catalase and magnesium porphyrins in chlorophylls. Isoprenoids are also ancient compounds, probably originally developed for protecting cellular components from photooxidation, and include ubiquinone (coenzyme Q), carotenoids and vitamin A (in rhodopsin).

2.4.1   FERMENTATION REACTIONS

The progenote presumably obtained energy for cell processes from fermentation reactions in which first aldehydes and ketones and later carbohydrates were partially oxidised anaerobically. These reactions were coupled to phosphorylation of ADP and involved the transfer of H atoms to organic molecules. These molecules became reduced and were excreted as organic acids (Fig. 2.6). In addition to providing available energy in the form of ATP, such fermentation reactions could also be developed *to provide reducing power in the form of NADH that could be used to drive biosynthetic reactions*. A most important form of biosynthesis was the fixation of CO_2 into organic compounds, probably initially pentoses like ribose, then hexoses like glucose, as well as a wide range of intermediates used in other synthetic pathways. The fixation of N_2 using nitrogenase enzymes was probably also an early development; this probably involved ferredoxin, as it does today in the supposedly ancient anaerobic *Clostridium*.

One result of the development of fermentation reactions would be the accumulation of acids tending to lower cytoplasmic pH and to impair the functioning of many enzymes. The early prokaryotes therefore needed to develop **proton pumps**, powered by ATP, to expel H^+ ions from the cell and raise internal pH (Fig. 2.7a). This use of ATP would be wasteful, especially if the supply of fermentable substrates was dwindling, and it is thought that the organisms developed transmembrane redox pumps to expel protons by coupling an external reaction involving an electron donor with an internal reaction involving an electron acceptor, the coupling being achieved by an electron-transport component (a uniquinone or something similar) in the membrane (e.g. Fig. 2.7b). Certain of these redox pumps can be so efficient in expelling protons that they set up a substantial electrochemical gradient (owing to high proton concentrations outside), sufficient to drive protons into the cell through pumps which could normally use ATP to expel protons. *Such inflow of protons drives the pump in reverse and generates ATP, providing additional energy for cellular activities (Fig. 2.7c).* Similar ATP synthetase systems driven by electrochemical proton gradients are widely used in prokaryotes, mitochondria and chloroplasts.

Figure 2.6 Some fermentation reactions producing ATP, and examples of prokaryotes which perform them.

glucose + ADP $\longrightarrow$ lactate + ATP   e.g. *Lactobacillus*

glucose + ADP $\longrightarrow$ ethanol + lactate + CO_2 + ATP   e.g. *Leuconostoc*

glucose + ADP $\longrightarrow$ ethanol + CO_2 + ATP   e.g. *Zymomonas*

lactate + ADP $\longrightarrow$ propionate + acetate + ATP + CO_2 e.g. *Clostridium*

2.4.2 USE OF LIGHT ENERGY

These more efficient fermentation processes still depended on the availability of organic molecules in the environment, and the supply of these was limited. There was an abundant source of carbon in CO_2, but strong reducing power in NADH (or NADPH) was required to assimilate it into organic compounds. The hydrogen plus electrons obtainable from readily available inorganic sources like H_2S or H_2O has too little reducing power without supplementary energy to transfer electrons (plus protons) to NAD^+, *but such energy could be obtained from a variety of chemical reactions or from light.* For example, anaerobic methanogens (archaebacteria) can obtain energy from the reduction of CO_2 to methane, by a pathway which involves fixation of CO_2 into organic compounds, and the eubacterium *Desulphovibrio* gains energy for CO_2 fixation by the reduction of $SO^2_4^-$ to H_2S, or by the activity of a hydrogenase enzyme in the production of hydrogen gas.

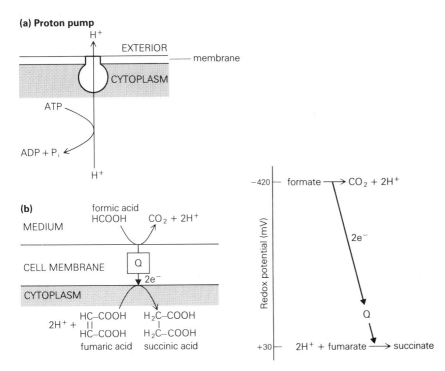

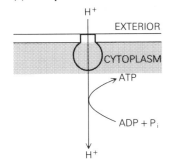

Figure 2.7 (a) A proton pump in the surface membrane of a bacterium can pump out protons to lower the internal pH of the cell. (b) A simple transmembrane redox pump found in the bacterium *Escherichia coli*. Its activity results in the removal of protons from the cell. The electron donor is coupled to the electron acceptor by a quinone-type carrier (Q) in the membrane. (c) In the presence of a substantial electrochemical gradient (pH or potential difference across the membrane), protons may be driven into the cytoplasm through the pump, which acts as an ATP synthetase.

Figure 2.8 The pigment bacteriorhodopsin, found in the surface membrane of some halobacteria, pumps protons out of the cell using absorbed light energy. The electrochemical proton gradient is used to generate ATP, as in Figure 2.7c.

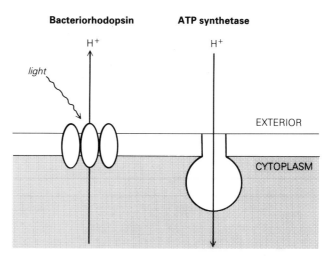

The surface membrane of some halobacteria (archaebacteria) contains bacteriorhodopsin which can absorb light energy. However, the light energy absorbed by bacteriorhodopsin is not used to generate reducing power but to drive an outward proton pump, analogous to the redox pump mentioned above, *thereby creating an electrochemical proton gradient from which an inward flux of protons generates ATP by ATP synthetase activity* (Fig. 2.8). Although they generate ATP for some cellular functions, such prokaryotes are still dependent on organic food in order to generate NADH for biosynthesis, and so are **photoheterotrophs** and not **photoautotrophs**.

Light of certain wavelengths between about 300 and 800 nm can be absorbed by various chlorophylls and can be used to energise electrons

Figure 2.9 Diagrams summarising two types of bacterial photosynthesis. In each of them light energy absorbed by bacteriochlorophyll that donates electrons (donor BChl) energises these electrons and they pass to an acceptor at a negative redox potential. In α the electrons return to the donor via carriers (ubiquinone uQ, cytochromes b and c), to complete a cyclical process that energises the intermediate Q, which in turn can either power phosphorylation of ADP or power the transfer of electrons from succinate to NAD$^+$. In b the bacteriochlorophyll obtains electrons from H$_2$S via coupled cytochromes which generate ATP, and pass electrons from the acceptor via carriers (ferredoxin fd and flavoprotein fp) to reduce NAD$^+$.

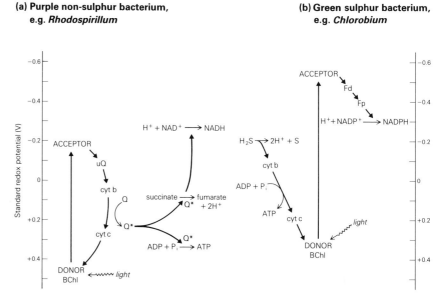

obtained when protons are removed from a hydrogen donor, which may be an organic compound, H_2S or H_2O. These electrons may be passed down an electron-transport chain, some stages of which may phosphorylate ADP to ATP, or, if they are strong enough electron donors, may reduce NAD^+ to NADH (or $NADP^+$ to NADPH). The ATP may be used in general cellular functions or may be used with NADH or NADPH in biosynthetic reactions, e.g. photosynthesis.

The bacteriochlorophylls of purple sulphur bacteria, green sulphur bacteria and purple non-sulphur bacteria are able to absorb sufficient light energy to energise electrons derived from organic compounds (Fig. 2.9a) or from H_2S (Fig. 2.9b) in the synthesis of ATP and/or NADH, but do not use water as a donor of electrons and hydrogen. This early form of photosynthesis was limited by the availability of suitable organic hydrogen donors or H_2S, and remains largely anaerobic. The cyanobacteria and chloroxybacteria combine chlorophyll a with light-absorbing phycobilic proteins or chlorophyll b, respectively, in coupled photosystems which absorb enough energy to use electrons derived from water in the synthesis of ATP and NADPH (Fig. 2.10), and release oxygen. The development of oxygenic photosynthesis by these prokaryotes, requiring only H_2O, CO_2 and light for organic synthesis, *at last gave its possessors freedom from such environmental constraints as the need*

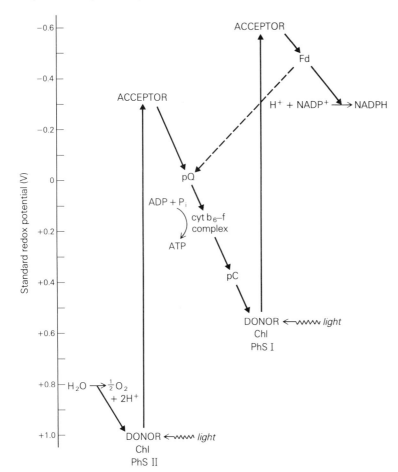

Figure 2.10 Coupled photosystems used by organisms containing chlorophyll a. Electrons obtained from water are energised by light energy absorbed in chlorophylls of photosystem II (Donor Chl PhS II); from the acceptor the electrons pass through carriers (plastoquinone pQ, cytochromes and plastocyanin pC, which participate in phosphorylation of ADP), to the chlorophylls of photosystem I, where they are again energised by light to reach a sufficiently negative redox potential to pass via ferredoxin to reduce $NADP^+$. Photosystem I can be operated cyclically to produce ATP (dotted line), and in this case NADPH is not produced or water split.

for H_2S, and led to the release of sufficient oxygen to produce an oxidising atmosphere.

The chlorophylls, enzymes and coenzymes of these photosystems and electron-transport chains are always associated with membranes and are normally integral components of membranes that form flattened sacs (**thylakoids**) within the cytoplasm. Components of the electron-transport chains pump protons across the thylakoid membranes into the thylakoid space to create an electrochemical gradient that drives the phosphorylation of ADP through an ATP synthetase system (Fig. 2.7c).

2.4.3   AEROBIC RESPIRATION

The cytochrome components of electron-transport chains developed for use in photosynthesis can also be reduced via NADH, etc., formed during breakdown of 2-carbon and 3-carbon molecules like acetate or pyruvate using Krebs Cycle enzymes, and can be used to generate abundant ATP if the terminal cytochrome c of the chain can be suitably oxidised. One solution used by anaerobically respiring heterotrophic bacteria is to oxidise cytochrome c by the reduction of nitrate to N_2, NO and H_2O. In the presence of free oxygen in the environment, O_2 is used as the final H acceptor in the cytochrome chain (Fig. 2.11). *The use of oxygen in this way by aerobic bacteria made possible a much greater efficiency of ATP synthesis using the Tricarboxyl Acid Cycle for the mobilisation of energy by breakdown of organic compounds to CO_2 and H_2O.*

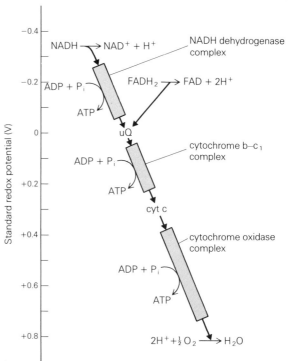

Figure 2.11 Redox diagram of the electron-transport chain used by aerobic bacteria and mitochondria to oxidise NADH and $FADH_2$ produced by the Krebs Cycle, showing steps that are associated with ATP synthesis.

Aerobic respiration was adopted by several lines of bacteria, notably those in lines containing chlorophyll a, the green filamentous bacteria, e.g. *Chloroflexus*, the purple non-sulphur bacteria, e.g. *Rhodospirillum*, and a diversity of Gram negative non-photosynthetic bacteria like *Paracoccus*, *Rhizobium* and *Agrobacterium* that appear on other criteria to have close affinities with the purple non-sulphur types. Thus *Rhodospirillum* can use its cytochrome chains for both photosynthesis and respiration, while the related *Paracoccus* appears to have lost photosynthetic ability and become totally heterotrophic using cytochromes only for respiration, with NO_3^- or O_2 as the hydrogen acceptor. The cytochromes in these non-photosynthetic forms remain as integral components of membranes, and are found either in the cell surface membrane or in invaginations of it called **mesosomes**. The possibility that cytochromes can be used to pass electrons to O_2 or to components of photosynthetic photosystems means that O_2 should if possible be kept away from photosynthetic reaction centres; indeed, high oxygen concentrations inhibit photosynthesis of cyanobacteria. Nitrogenase enzymes used in fixation of nitrogen are also inhibited by oxygen, and those cyanobacteria that fix nitrogen have thick-walled **heterocysts** with special wall lipids that reduce oxygen levels within the cell; oxygen levels are also kept low by intense respiratory activity.

The availability of atmospheric oxygen provided various pathways for aerobic bacteria to gain energy by inorganic oxidation reactions. Thus *Hydrogenomonas* uses O_2 to oxidise molecular hydrogen obtained from the environment, and can live as a **chemoautotroph** by fixing CO_2, although it uses organic carbon sources if they are available. Several bacteria can derive energy from the aerobic oxidation of inorganic nitrogen compounds; e.g. *Nitrosomonas* which oxidise NH_4^+ to NO_2^- and water, and *Nitrobacter* which oxidises NO_2^- to NO_3^-. These can live as true chemoautotrophs (chemolithotrophs), and are essential organisms in the nitrogen cycle, but can also use organic carbon sources. *Methylomonas* oxidises methane or methanol, and is important in the recycling of these highly reduced carbon compounds. Bacteria that cycle sulphur compounds include *Thiobacillus*, species of which oxidise sulphide, thiosulphate or sulphite to sulphate, and *Sulpholobus* (an archaebacterium inhabiting hot springs) which oxidises elemental sulphur. *Thiobacillus ferrooxidans* obtains energy by promoting the oxidation of the Fe^{2+} ion to Fe^{3+}.

The patterns of energy metabolism outlined in these pages suggest that the evolution of prokaryotic cells *depended upon the development within the cells of molecules that could transfer energy (as electrons) and others that could energise electrons using light energy, as well as upon the availability of such chemical components as organic compounds, H_2S and O_2 from the environment.* Possible evolutionary pathways involving these organisms are shown in Figure 2.12.

2.5 Cellular features of prokaryotes

It was commented earlier that structural features show little development among prokaryotes. Some membranous structures, the thylakoids and mesosomes, primitively associated with the cell membrane, are developed to accommodate photosynthetic pigments and coenzymes of electron-transport

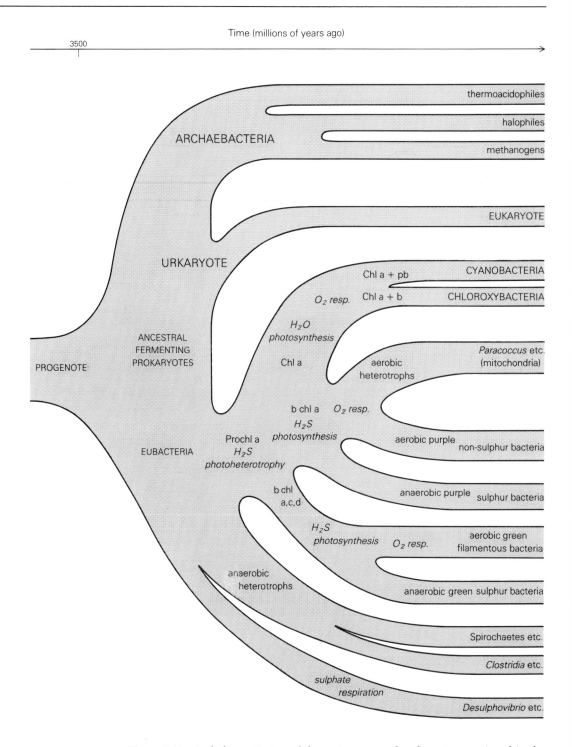

Figure 2.12 A phylogenetic tree of the main groups of prokaryotes mentioned in the text. Ch1 is chlorophyll; bch1, bacteriochlorophyll; pb, **phycobilins** Prochl a, protochlorophyll a and Resp. is respiration.

chains, but other membranous structures are rare. However, features of the cell wall, capsule and appendages are important characteristics of various prokaryotes (Fig. 2.13). Characteristics of the prokaryotic chromosomal apparatus are also important features.

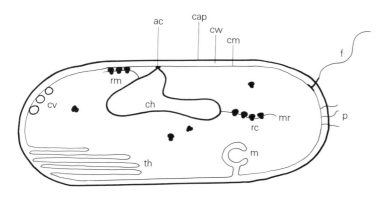

Figure 2.13 A diagram of structures seen in prokaryote cells: ac, attachment of chromosome to membrane; cap, capsule; ch, chromosome; cm, cell membrane; cv, chlorobium vesicles (in green sulphur bacteria); cw, cell wall; f, flagellum; m, mesosome; mr, messenger RNA; p, pili; rc, ribosomes producing cytoplasmic protein; rm, ribosomes attached to cell membrane (and mRNA); th, thylakoid (in purple photosynthetic bacteria).

2.5.1 CELL WALLS AND FLAGELLA

The majority of bacteria secrete a cell wall outside the cell membrane. The main structural constituent of the eubacterial cell wall is a *peptidoglycan* (**mucopeptide**), muramic acid. Bacteria are grouped into two main classes, Gram positive and Gram negative, according to the ability of the cell to retain Gram's stain (crystal violet) after washing with a suitable organic solvent. This differentiation is thought to be based upon different characteristics of the cell wall, which in Gram positive bacteria may have up to 95 per cent of mucopeptides while in Gram negative species may have as little as 5 per cent. In Gram negative forms there are substantial amounts of phospholipid and lipopolysaccharide in a cell wall that appears to consist of an outer membrane separated by a space from the cell membrane, and is associated with a coat of lipopolysaccharides that anchor proteins and carbohydrates of the surrounding capsule. Cyanobacteria have cell envelopes similar to those of Gram negative bacteria. Gram positive bacterial cell walls are more amorphous and contain teichoic acid associated with the mucopeptide. Capsule (sheath or slime) layers outside the cell wall are extremely variable in extent and in composition; they are usually largely polysaccharide in nature, but polypeptides and other compounds may also be present. Some bacteria carry filamentous appendages called **pili**, usually less than 10 nm thick and less than 1 μm long, attached to the surface of the cell wall or capsule; tubular 'sex pili' are borne by some bacteria (see below). Bacteria of various groups may, under certain conditions, lack cell walls. These are the so-called L-forms and they are structurally similar to bacteria like *Mycoplasma* which permanently lack cell walls.

Rigid helical flagella, 20 nm thick and several micrometres long, are attached to the membranes of some bacteria. The *structure, chemistry and motility of bacterial flagella are all different from these features of eukaryote flagella* (see §7.5 & §7.6). Such flagella may be present all over the cell, or may be limited to one or both poles of the cell, or only a single flagellum may be present. Neither cyanobacteria nor chloroxybacteria possess flagella, but some cyanobacteria and various other bacteria show gliding motility. Some

large spirochaetes contain internal fibrils that look like **microtubules** (see
§7.4) and are composed of a protein similar to **tubulin**; there is no evidence
that they are involved in motility.

2.5.2 PROKARYOTIC CHROMOSOMES

The prokaryote genome has been studied in several eubacteria, including
cyanobacteria, and always seems to take the *form of a single double-stranded
helix of DNA formed into a closed loop*. It is permanently attached to the cell
membrane at least at one point, and when isolated often carries a number of
membrane components with it. Membrane-free preparations of bacterial
chromosomes contain small amounts of protein, principally RNA polymerase
enzyme, although small quantities of a small heat-stable (HU) protein that
may *be analogous to eukaryote histones have been found in some species*.
The many-looped form of bacterial DNA seems to be maintained by RNA, but
it is not clear what function this serves (see §5.4.1).

All three classes of RNA (mRNA, tRNA and rRNA) are formed by activity of
the same RNA polymerase in prokaryotes. The messenger RNA formed at the
chromosome is directly available for translation without processing, and so
ribosomes may attach to the beginning of the mRNA strand and commence
translation while the other end of the mRNA is still being formed by
transcription from DNA. Proteins for use within the cell are synthesised at
cytoplasmic ribosomes, but ribosomes responsible for the synthesis of
membrane proteins or proteins destined for export from the cell to form the
cell wall or secretory products are attached to the cell membrane and the
resulting polypeptides are ejected directly into or through the membrane as
they are formed; the DNA may thus be connected to the cell membrane at
additional points through mRNA molecules (Fig. 2.13). The structure of
prokaryotic ribosomes is described in Section 6.2.1.

In addition to the chromosome most bacteria contain shorter lengths of
DNA called **plasmids**, which also form closed loops. A particular cell may
carry single or multiple copies of one or several plasmids. Some plasmids are
merely **bacteriophage (viral) DNA**, which may alternatively be incorporated
within the chromosome (see §5.6). Other plasmids may be separated parts of
the normal genome from the same or a foreign cell, and may recombine with
the main chromosome.

DNA replication in prokaryotes is normally continuous throughout the cell
cycle. The replication of bacterial chromosomes is discussed in Section 5.3.1.
Bacteria participate in various forms of gene transfer and recombination, even
between different species, none of which is quite the same as sex in
eukaryotes, but which have implications for both the evolution of prokaryotes
and gene cloning (see §5.6).

Organisms with these characteristics *dominated life on Earth for a major
part of its history. They came to exploit all environments available for life,
including some not entered by any eukaryote*. Because of their cellular
design, they have limited size, though colonial forms like actinomycetes and
some cyanobacteria produce sizeable aggregates, and cyanobacteria show
some cell differentiation. The evolution of the more efficient eukaryotes
eventually led to the overshadowing of prokaryotes in many major habitats,
yet the prokaryotes remain extremely important through their roles in element
cycling and microbial disease.

2.6 Cell organisation in eukaryotes

2.6.1 NUCLEI

The prime distinguishing feature of eukaryotes is the possession of a *membrane-bounded nucleus with multiple linear chromosomes* in place of the prokaryote nucleoid region with a single circular chromosome. Each linear eukaryote chromosome contains a single twin-stranded helix of DNA which is associated with sets of five types of **histone proteins** to form **nucleosomes** about 10 nm in diameter that appear like beads on a string in uncondensed (interphase) chromosomes (see §5.4.1). Replication of chromosomes is restricted to a certain period of the cell cycle, called the **S-phase**, and is accompanied by a massive synthesis of histone proteins and the formation of new nucleosomes. The result of these processes is that each chromosome consists of two chromatids that remain joined together at the centromere until mitosis.

The DNA provides templates for transcription in synthesis of RNA, which, in eukaryotes, involves activity of three different RNA polymerases, one for mRNA, one for large rRNA and the third for small tRNA and rRNA molecules (see §6.3 & Table 6.1). The total DNA in a eukaryote cell is very much greater than in a prokaryote cell. It is in fact greater than one would expect even given that there are more functional genes in the more complicated eukaryotic cell (see §5.4 and Table 5.3).

The nuclear envelope is formed of two unit membranes separated by a perinuclear space 30 nm or more thick. The outer membrane is continuous with endoplasmic reticulum (ER) membranes, and the perinuclear space is continuous with ER cisternae (Fig. 2.14). *The presence of the nuclear*

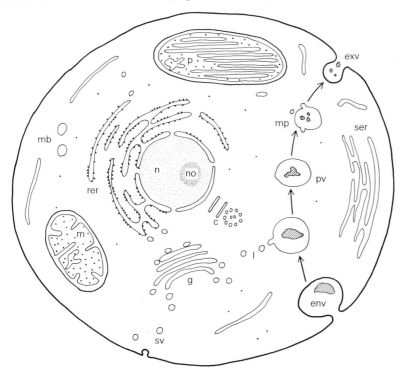

Figure 2.14 Membranous structures seen in eukaryote cells: env, endocytotic vacuole; exv, exocytosis of vacuole; g, Golgi apparatus; l, lysosomes; m, mitochondrion; mb, microbodies; mp, micropinocytotic vesicles; n, nucleus; p, plastid; pv, phagocytic vacuole; rer, rough endoplasmic reticulum; ser, smooth endoplasmic reticulum; sv, secretory vesicle. Centrioles (c) and nucleolus (no) are also shown.

envelope in eukaryotes is probably essential to separate transcription from translation, because of the RNA processing that intervenes; no such separation is required in prokaryotes where this RNA processing does not occur. A detailed description of the nucleus is given in Section 5.5.

Nuclear division must follow the DNA replication of the S-phase, but must precede cell division. Each daughter cell must receive a complete set of chromosomes from the parent cell; this is the function of mitosis, which is necessarily more complex than the separation of chromosomes in prokaryote division because of the presence of multiple chromosomes. It would be possible to separate daughter replicated chromosomes (**chromatids**) by migration of duplicated centromeres attached to the nuclear envelope as two groups to opposite poles of the nucleus, and there is indeed a remnant of such a mechanism in some eukaryotes. However, this evidently proved inadequate to ensure the separation of complete chromosome sets, so it was first supplemented by and then replaced by the nuclear spindle of **microtubular fibrils** in the performance of this role. The microtubules develop from **microtubule organising centres (MTOCs)**, which in this case are usually **centrioles** or other spindle pole bodies, at either pole of the nucleus (see Figs 7.19-7.22).

A detailed account of chromosome movements during mitosis is given in Section 7.4.4.

Most eukaryotes show the phenomenon of sex. Usually this involves a meeting of two gamete cells with a single (haploid) set of chromosomes and fusion of their nuclei to make a single (diploid) **synkaryon** with two chromosomes of each type. Such a diploid cell may immediately undergo a special meiotic division which reduces the chromosomes to the haploid number, or may grow and divide mitotically for a number of cell generations before undergoing meiosis so that the progeny can once again participate in **syngamy**. The meiotic process is described in detail in another volume in Modern Views in Biology (*Reproductive Biology*), (Ch. 1) but typically involves two divisions of the nucleus in which movements of the chromosomes are controlled by spindle microtubules. *Genetic variation within the population is much increased because new chromosome combinations form during syngamy and because during meiosis chromosomes segregate independently and genes recombine through crossing over.*

2.6.2   MEMBRANE SYSTEMS

Compartmentation within eukaryote cells is provided by internal membranes (Fig. 2.14). One example of this has been seen in the nuclear envelope, which not only separates nucleoplasm from cytoplasm, but also provides, between the two membranes of the nuclear envelope, an intermembrane space that is continuous with the space within elements of the cytoplasmic membrane system of ER. This ER is composed of membranous sacs or channels, which are believed to be extensively interconnected, and in a state of dynamic flux, participating in fusions and separations of its constituent elements. Some membranes of the system (rough ER) may carry ribosomes on their cytoplasmic face, and such ribosomes may be involved in synthesis of polypeptides that are inserted into the membrane or passed directly into the cisternal space within the ER element for processing before use or export from the cell, usually via the **Golgi apparatus**, a special region of the ER (see §4.2.1). Other parts of the system lack ribosomes (smooth ER) and serve other functions, e.g. as a site for synthesis of membrane lipids. The ER membrane system also produces microbodies (peroxysomes, glyoxysomes, etc.).

Membranes are involved in the export of material by **exocytosis** (secretion) and import during **endocytosis** (phagocytosis or pinocytosis) (see §4.2.1), which are all eukaryote features. In eukaryotes imported materials are usually broken down within the cell in phagocytic vacuoles by the action of hydrolytic enzymes from **lysosomes**, rather than being digested outside the cell by secreted enzymes as in prokaryotes. Lysosomes are small vesicles produced at the Golgi apparatus, and contain a diversity of acid hydrolases.

2.6.3 MITOCHONDRIA AND CHLOROPLASTS

Mitochondria and chloroplasts are cytoplasmic 'energy organelles' bounded by two membranes and containing their own DNA and protein synthesis systems. The DNA found in the lumen of organelles of both types occurs as *small, normally circular, molecules, probably attached to the inner membrane*. The ribosomes present in the organelles are usually smaller than those found in the cytoplasm, and are sensitive to antibiotics that act upon bacterial protein synthesis rather than those that affect activity of eukaryote ribosomes; chloroplast ribosomes in particular are similar to the 70 S ribosomes of such bacteria as *E. coli*, while mitochondrial ribosomes have rather more variable features. The organelle DNA codes for organellar rRNA and tRNAs, but for mRNAs of only a few of the organellar proteins. *Most of the proteins functional in the organelles are made in the cell cytoplasm from nuclear mRNA and are actively transported into the organelles*; these include, for example, the mitochondrial RNA polymerase, so all transcription in mitochondria is dependent upon an enzyme encoded in nuclear DNA.

Mitochondria are generally rod-shaped or filamentous organelles about 0.5-1.0 μm in diameter that show dynamic shape changes. The outer membrane is normally smooth, but the inner one is extensively folded inwards to form **cristae**. These two membranes have different compositions; in particular, the outer membrane is relatively permeable to small molecules, while the inner membrane is much less permeable and contains integral proteins of the electron-transfer chain, ATP synthetase and a number of specific transport enzymes. The lumen (matrix) contains numerous enzymes, notably those concerned with the Tricarboxylic Acid Cycle (TCA) which breaks down pyruvate and fatty acids brought into the mitochondrion from the cytoplasm and produces NADH that passes electrons to the electron-transport chains in the membrane. The space between the two membranes forms a compartment that is *usually at a more acid pH than the lumen within the inner membrane because protons are released into the intermembrane space by activity of the electron-transport chains*. ATP is formed in the lumen by activity of ATP synthetase driven by the proton gradient between the intermembrane space and the lumen (see Fig. 2.7c). A special transport mechanism makes this ATP available for energy-requiring reactions in the cell cytoplasm and nucleus as well as within the mitochondrion.

Chloroplasts are usually ovoid or lenticular organelles 2-5 μm across whose two outer membranes enclose lumenal cytoplasm (**stroma**) containing **thylakoids** (flattened membranous sacs) that may be separate or aggregated in various ways. The outer membrane is more permeable than the inner one, which contains some transport proteins, but this inner membrane remains parallel to the outer one, without cristae. Electron-transport chain complexes and ATP synthetase are not located in the inner chloroplast membrane but in the thylakoid membranes, where they accompany chlorophyll and associated light-absorbing molecules. The photosynthetic electron-transport chains generate a proton gradient between the intrathylakoid space (at acid pH) and the stroma, *within which ATP is generated by ATP synthetase enzymes in the thylakoid membrane, driven by the proton gradient.* NADPH is also generated in the stroma by **photosystem I** of photosynthesis. The carbon fixation reactions of the chloroplast use ATP and NADPH produced by photosystems I and II in the synthesis of carbohydrate catalysed by enzymes in the stroma; synthesis of fatty acids, amino acids and nucleotides also occurs in the stroma, using the same energy sources. Synthesised products, particularly glyceraldehyde-3-phosphate, are transported out of the chloroplast to provide molecules for energy production and synthesis in the cytoplasm. Synthetic products may also be elaborated into food storage bodies of carbohydrate or lipid within the chloroplast or cytoplasm.

2.6.4 CYTOSKELETON AND FIBRE SYSTEMS

The structurally complex cells of eukaryotes, with their transport systems and other motile properties, possess proteinaceous fibres that provide a cyto-skeletal framework for the positioning of cell organelles and maintenance of cell shape as well as elements of a diversity of motile systems. Some cytoskeletons are dynamic, and others are static, depending on which of the three principal classes of fibrous elements predominate; **microfilaments**, **microtubules** and **intermediate filaments** are *all present in most eukaryote cells, but such fibres are rarely if ever present in prokaryotes.* The structures and functions of these three classes of fibres are described in Section 7.1.

2.6.5 CELL SURFACE SPECIALISATIONS

Eukaryotes show developments of the cell surface for protection or motility just as prokaryotes do, though the details differ, and because of the complex multicellular natures of many eukaryotes, special junctional structures are found. Cell walls whose components are secreted through the surfaces of some eukaryote cells are usually largely polysaccharide in nature, based on either cellulose or chitin. Silica structures produced by some protists as cell walls or surface scales are secreted within Golgi vesicles and appear later at the cell surface. Many organisms secrete polypeptides that assume a structural role as, for example, a collagen matrix between cells or as a basis for impregnation by calcium salts in shells or some other skeletons. Surface polypeptides, especially glycoproteins, may form part of a protective cell coat or layer of scales on otherwise naked cells. *Muramic acid, found in the cell walls of all eubacteria, is never present in eukaryotes.*

Cells without rigid surface protection may have a specialised cortical region, the most superficial part of which may form a pellicle consisting of the surface membrane and layers immediately inside it. The structures involved may be cytoskeletal fibrils, particularly microtubules, which may be interspersed with dense granular cytoplasm, and in some protists a layer of vesicles lines the cell surface.

The presence of surface specialisations limits the extent of endocytosis and exocytosis. Cells with a naked surface may be capable of uptake of materials by vesicle formation, or of release of vesicle contents, at any part of the cell surface. *Cells with cell walls normally show neither endocytosis nor exocytosis, though some secreted products may be released from vesicles at the cell surface during wall formation.* Where the cell secretes a partial shell or develops a pellicle, specific sites of phagocytosis and of release of materials occur, e.g. at a **cytostome** (cell mouth), **cytopyge** (cell anus) or contractile vacuole pore.

In multicellular systems, special junctional complexes involving the membranes of adjacent cells are developed with specific functions. Cell junctions are described in detail in Section 4.3.

Flagellar projections from eukaryote cells are not rigid extracellular filaments like bacterial flagella, but are extensions of the cell membrane, usually 2-20 μm long and 0.25 μm in diameter, containing a cylindrical bundle of 20 microtubules (the **axoneme**) arranged as 9 outer doublets and 2 single central microtubules. This is often referred to as the 9+2 pattern (see §7.5).

2.7   The origin of eukaryotes

In the previous sections we have seen that eukaryotes differ from prokaryotes in many important ways, involving both nuclear and cytoplasmic components. The more complex nucleus in eukaryotes is enclosed by membranes and includes several or many linear chromosomes whose separation after division involves a microtubular spindle, and from which RNA is transcribed by activity of several RNA polymerases, all RNAs being subsequently processed in the nucleus before they are functional. The more complex cytoplasm of eukaryotes contains larger ribosomes than in prokaryotes and has various fibrous structures (including 9 + 2 flagella), endoplasmic reticulum and Golgi and other vesicles, including those involved in both phagocytosis (endocytosis) and secretion (exocytosis), as well as usually containing mitochondria and often containing chloroplasts.

Stress has often been placed on the significance of mitochondria and chloroplasts within cells as indicators of the transition between prokaryotes and eukaryotes. There is considerable support, discussed below, for the view that both *mitochondria and chloroplasts are descended from symbiotic prokaryotes living within the eukaryote cell*. Clearly, however, although the presence of mitochondria (and sometimes chloroplasts) may have given the eukaryotes such an evolutionary advantage that their possessors have come to characterise the eukaryote lineage, it is also true that all eukaryotes also possess many other non-prokaryote features, and the *ancestral eukaryote may be assumed to have possessed most, if not all, of these before they acquired mitochondrial and chloroplast symbionts*. Indeed, there are several groups of protists that have neither mitochondria nor chloroplasts, and one of these, the Parabasalia, contains flagellates with complex eukaryotic cytoplasmic structures and primitive nuclear features: these could have a direct lineage from eukaryotes that never acquired mitochondria, although many of them are parasites and may have less need of mitochondria.

The origin of the typical eukaryote pattern therefore seems to have involved two main stages: first the development of the nuclear and cytoplasmic organisation typical of an anaerobic phagotrophic eukaryote, and second the acquisition by this anaerobic form of an aerobic prokaryotic symbiont for the efficient production of ATP from organic food and in some cases also of a chlorophyll-containing symbiont for trapping light energy for organic synthesis.

It was mentioned earlier that quite fundamental differences occur between archaebacteria, eubacteria and eukaryotes, and the basic nature of the differences in ribosomal RNA sequences and functioning (among others) suggest that the two bacterial groups and the urkaryote separated at a very early stage in the evolution of prokaryotes, *after the development of the genetic code of DNA and energy transduction by ATP/NAD systems, but while other cellular biochemical features were still varied.* The urkaryote is thought to have evolved in parallel with, but separate from, the bacterial groups (Fig. 2.12), and to have progressively developed the various nuclear and cytoplasmic eukaryote features. Major steps in this process would involve the development of endocytosis, with associated elaborations of membrane and fibrillar systems within the cell to process and transport phagocytic vacuoles, and of the nuclear apparatus, mitosis, sex and meiosis. Phago-trophic nutrition would set such organisms apart from any known bacteria, giving them a different ecological role and encouraging the development of both amoeboid and flagellate features for the more efficient capture of bacteria or organic debris for food.

Such early eukaryotes would have been limited to relatively inefficient anaerobic energy-producing processes, and may only have been able to survive in poorly oxygenated places where food was abundant. However, when some oxygen-tolerant eukaryotes entered into successful symbiosis with certain aerobic bacteria and enhanced their efficiency of energy production as well as needing access to oxygen, *they would have been able to exploit more habitats and outcompete most anaerobic eukaryotes in a world of progres-sively increasing oxygen concentrations.* Few, if any, traces of these original anaerobic eukaryotes therefore remain. Some aerobic eukaryotes proceeded to acquire photosynthetic **symbionts** of one type or another and became successful autotrophic eukaryotes. In time the heterotrophic and autotrophic eukaryotes replaced prokaryotes as the dominant life forms on Earth, and, while the prokaryotes remained abundant in a number of ecological situations that eukaryotes have not successfully exploited, the eukaryote cell pattern evolved extensively, first in the various groups of protist and later in the major lines of fungi, animals and higher plants, to occupy and dominate most ecosystems.

2.8   Symbiotic organelles in eukaryotes

The presence of prokaryotic or even eukaryotic symbionts within the cells of such phagotrophic eukaryotes as amoebas or ciliates is extremely common, showing that *symbiotic associations are readily formed and successfully maintained.* It seems likely that ever since the early eukaryote became phagotrophic it has been possible for food organisms to become symbiotic by somehow avoiding digestion within endocytic vacuoles and establishing mutually beneficial chemical exchanges with the host cell cytoplasm. *Both mitochondria and chloroplasts may have entered eukaryote cells in this way, and gained recognition so that they were not digested but became permanent inhabitants.* If this is true, the outer membrane of the organelle would be derived from the host cell vacuole and the inner membrane from the symbiont; there is indeed evidence not only that the two membranes are different, but also that the *outer membrane is more like host cell ER membrane, while the inner membrane has prokaryotic features.* Both types of organelle contain circular DNA without histones, and also usually contain smaller 60 S or 70 S ribosomes capable of translating mRNA derived from organellar DNA. They both maintain complete genetic continuity by division and by mechanisms which ensure their presence in both daughters during division of the 'host' eukaryote cell. Comparisons of mitochondrial biochemistry with that of the bacterium *Paracoccus* reveal an identical electron-transport system with the same cytochromes and the same behaviour with antibiotics. The photosynthetic apparatus in chloroplasts is somewhat different in different eukaryote groups, but *the thylakoid organisation and pigment composition of chloroplasts in red algae appears identical to these features in cyanobacteria.* Analysis of ribosomal RNA shows that mitochondrial sequences are more similar to bacterial (and chloroplast) sequences than to those from eukaryotic cytoplasmic ribosomes.

Such evidence has led to the wide acceptance of the view that mitochondria were derived from symbiotic prokaryotes and almost universal acceptance of a symbiotic origin for chloroplasts. This does not mean that mitochondria of all eukaryotes were necessarily derived from the same symbiotic event; *the early eukaryotes may already have evolved different cell patterns before acquiring mitochondrial symbionts,* which themselves may have been somewhat different, so we need not expect mitochondria in all eukaryote groups to be identical. The diversity of chloroplasts is even more marked, and strongly indicates that different eukaryotes entered into permanent symbiotic relationships with different autotrophs, some prokaryotic, some eukaryotic. It is assumed that for both symbiotic organelles there has been a progressive transfer of genetic responsibility for organellar proteins to the host cell nucleus; this transfer has proceeded to different extents in different groups of eukaryotes. There has also been some reduction in the number of tRNA molecules and some have become aberrant in structure; in consequence *the reading of the genetic code by tRNAs has become variable in mitochondria* (see §6.7.2). These developments have made the symbiont progressively more dependent upon the host cell, just as the eukaryote has become dependent upon its organellar symbiont.

The alternative to a symbiotic origin for these organelles would be the progressive development within eukaryotic cells of membranes to isolate regions of cytoplasm enclosing a (plasmid?) genetic/protein synthesis mechanism responsible for the production of the appropriate biochemical systems. Such isolation would have to have been an ancient event, because of *the prokaryotic nature of the genetic/protein synthesis mechanisms,* and must have *protected the organellar DNA from factors that modified nuclear DNA.* The aerobic requirements of mitochondria mean that such organelles would have no value until oxygen was available in the environment, which was probably a relatively late event in prokaryote diversification. Support for such alternative explanations has rapidly waned in recent years as more information about organellar biochemistry has accumulated, although some scientists still believe that the evidence for the symbiotic origin of mitochondria is inadequate.

It has also been suggested that eukaryote microtubule systems, or at least those responsible for the production of 9 + 2 ciliary organelles, had an origin from symbiotic prokaryotes. This hypothesis suggests that ectosymbiotic spirochaetes (or spiroplasms) containing microtubular fibrils provided motility for early eukaryotes (as they do for some protists today), and that such spirochaetes became integrated into the eukaryote cell so that genes derived from spirochaetes were used in eukaryote descendants for the production of microtubules, microtubular structures such as centrioles and other microtubule-organising centres. Mitosis and meiosis could not therefore develop in their present form until microtubule components derived from symbiont genes became available within the cell.

This hypothesis has not found general acceptance because of the lack of substantial supporting evidence. In no prokaryote, as far as we know, *are microtubules used as a means of motility, and there is even doubt about whether fibres that resemble microtubules in spirochaetes and spiroplasms are actually composed of tubulin.* There is no evidence of any prokaryotic features associated with any of the structures concerned. It was once thought that DNA was associated with ciliary basal bodies, but this has now been disproved; there is some evidence that basal bodies contain RNA. There are also now strong doubts about the universal cytoplasmic inheritance in eukaryotes of structures acting as microtubule-organising centres (centrioles, etc.), which was formerly assumed.

The favoured present view is that microtubules developed in the early eukaryote as a means of maintaining cell shape in the absence of cell walls, or for producing changes in cell shape or in the position of cell structures. Such microtubules later became involved in the mitotic separation of chromosomes and as bundles in the axonemes of cilia and flagella. For each of these functions a number of ancillary proteins were progressively adapted or developed. While mitotic spindle microtubules appear to be present in all present-day eukaryote groups, several groups, including red algae and true fungi, lack 9 + 2 flagella. It is possible that the development of first the mitotic spindle and then 9 + 2 flagella preceded mitochondrial symbiosis – so that some eukaryote groups in which mitosis is present but not flagella must be assumed to have lost the latter. It seems more likely that the early eukaryotes had diversified into flagellate and non-flagellate groups before they acquired mitochondria, or that they evolved flagella after acquisition of mitochondria, so that present-day groups that lack flagella may never have possessed them.

2.9 Some variants of cellular features in eukaryotes

Eukaryote cells with most or all of the characteristics discussed in the previous sections were involved in an extensive adaptive radiation at the cellular level that produced a large number of different cellular patterns recognised as the various major taxa of protists, of which some authorities identify up to 30 or 40 phyla living today, as well as lines that led to fungi, green land plants and higher animals. A full description of cellular evolution in all of these lines is clearly beyond the scope of this account. It is proposed therefore to consider the major variants of the principal cellular features that characterise the diversity of eukaryotes, and discuss briefly their supposed evolutionary relationships.

2.9.1 NUCLEI AND NUCLEAR DIVISION

Details of chromosomal structures are available for only some of the groups of protists. In most eukaryote groups the DNA strand is complexed with histones to form conventional nucleosomes and the chromosomes are extended during interphase and condensed only during mitosis (see §5.4.1). In several groups the chromosomes remain condensed throughout interphase to a greater or lesser extent, and in some cases chromosomes appear not to condense during mitosis. Persistent condensation of chromosomes is most marked in dinoflagellates where there appears to be only a *single histone-like protein and nucleosomes are not formed*. Although protists of most groups, as well as higher eukaryotes, possess a full complement of histones, the individual histone proteins are variable. Cell size in protists is extremely variable and is often much larger than in higher animals and plants; *in most types of larger cells the number of gene sets within the cell is increased to cope with the greater cytoplasmic volumes by formation of polyploid nuclei or multi-nucleate cells, or both.*

Nuclear division has been studied more extensively and varies in many ways (Fig. 2.15). In several groups the nuclear membrane remains intact throughout mitosis (**closed mitosis**), in many others it disperses early in mitosis and re-forms afterwards (**open mitosis**), and in the remainder the membrane breaks down to a greater or lesser extent at the spindle poles, persisting equatorially, or only disperses late in mitosis. The mitotic spindle may be formed entirely outside or entirely inside a closed nucleus, but passes right through the nuclear area in open mitosis. The MTOCs that occupy the poles of the spindle may be centrioles, flagellar basal bodies (or structures attached to them by rootlets), plaques of material with no defined substructure, or in some cases no polar structures have been found. The extent of development of **kinetochores** is variable, but because there is

Figure 2.15 Some variants of mitosis shown during anaphase (see text): a, attractophore; c, centrioles; ne, nuclear envelope; no, nucleolus; p, plaque; po, polar opening in nuclear envelope.

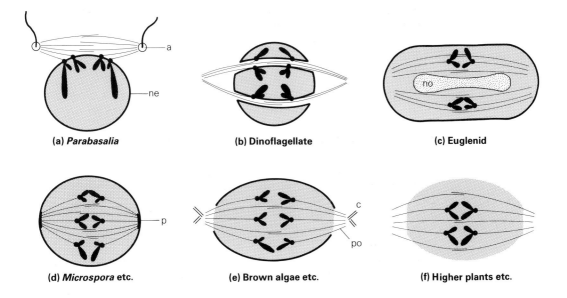

(a) *Parabasalia*   (b) Dinoflagellate   (c) Euglenid

(d) *Microspora* etc.   (e) Brown algae etc.   (f) Higher plants etc.

inadequate information for many groups, comparison is probably not worth while at present.

During the closed mitosis of parabasalian flagellates (*Trichomonas*, *Trichonympha*, etc.) the chromosomes remain attached at their centromeres to the inner side of the nuclear membrane throughout mitosis (Fig. 2.15a). The flagellar apparatus of a cell undergoes replication before mitosis and the two flagellar centres move apart; the mitotic spindle forms across one side of the nucleus between **attractophores** (MTOCs on short rootlets from special flagellar bases). The kinetochores of chromatids project through the nuclear envelope and the kinetochore microtubules therefore attach to the outer surface of the nuclear envelope at sites where the chromatids are attached on the inside. As the flagellar bases move apart the two groups of chromatids separate towards the sides of the closed nucleus with the interpolation of membrane between separating kinetochores. There have been suggestions of separation of kinetochores (by membrane growth?) before attachment of spindle microtubules, but this is not easy to prove.

In the closed mitosis of dinoflagellates the kinetochores again project through the nuclear envelope, but in this case the nucleus becomes modified early in mitosis by the formation of cytoplasmic channels lined by nuclear envelopes that run through the nucleus (Fig. 2.15b). Both polar and kinetochore microtubules of the spindle run through the channels, and the chromosomes lie in nucleoplasm around the channels with their kinetochores projecting through the channel walls to make contact with kinetochore microtubules in the channels. In most dinoflagellates there are no obvious spindle pole bodies and many channels run through the nucleus, but in one parasitic subgroup there is a single nuclear channel and pairs of centrioles occupy the spindle poles.

Closed mitosis is also found in many other eukaryote groups, but in these cases the microtubules of the spindle form entirely within the nucleus. In euglenid flagellates the flagellar basal structures duplicate and move apart, and the nucleus moves forward to lie betwen the two pairs of flagellar bases. A bundle of microtubules develops across the nucleus between the pairs of basal bodies, but the microtubules are intranuclear and no spindle pole bodies are seen – the microtubules apparently have no connection to the nuclear envelope or to the flagellar bases outside it (Fig. 2.15c). Some microtubules attach to kinetochores, and the chromosomes separate into two groups during elongation of the nucleus. The nucleolus does not disappear during mitosis, but extends and breaks in two as the nucleus elongates. There are no spindle pole bodies in either trypanosomes or in many ciliate micronuclei, and kinetochores appear to be absent in some instances. In trypanosomes and the amoeboflagellate *Naegleria* (whose chromosomes remain uncondensed) it appears that growth of the nuclear envelope, to which chromosomes remain attached, may be more important for separation of chromosomes than spindle activity. Spindle pole bodies are present as plaques in the nuclear envelope, or attached to its inner or outer surface, during the closed mitosis of ascomycete fungi, microsporans, some green and colourless flagellates, some ciliates and some sporozoans (Fig. 2.15d). In other sporozoans, radiolarians, oomycetes and members of some algal groups these spindle pole bodies may be associated with or replaced by pairs of centrioles lying in the cytoplasm immediately outside the closed nuclear envelope. Ciliate macronuclear chromatin remains attached to the inner nuclear membrane throughout '**amitotic**' division, during which neither an organised spindle nor condensed chromosomes are visible.

Polar openings or fenestrae are formed in the nuclear envelope during mitosis of representatives of other groups (Fig. 2.15e). The spindle micro-tubules that pass through the nucleus may be associated with pairs of centrioles (or basal bodies) in brown algae, some chytrids and some green algae, with other forms of spindle pole body in the cytoplasm of some amoebae and red algae, or polar structures may be absent in some other amoeboid forms. The extent of breakdown of the nuclear envelope may increase as mitosis progresses; some forms that start with closed mitosis may show breakdown or replacement of the nuclear envelope in the final stages of mitosis. Complete dispersal of the nuclear envelope early in mitosis is found in various amoeboid groups, some green algae, cryptophytes, chrysophytes, diatoms, metaphyta and metazoa (Fig. 2.15f). The poles of the spindle may be occupied by centrioles (the metazoan pattern) or other spindle pole bodies, but in some cases, as in metaphytes, no structures can be found at the poles.

2.9.2 LIFE CYCLES AND SEXUAL PROCESSES

After a eukaryote cell has undergone nuclear division by the processes described above, division of the cell normally follows, with a distribution of essential organelles, such as mitochondria, chloroplasts and centrioles, into both daughter cells; rarely, cell division is delayed and multinucleate cells are formed. Following cell division the cells of most protists separate to follow an independent unicellular existence, *but in most groups there are colonial representatives whose cells do not separate, and in several groups quite complex multicellular organisms result.* Thus the red and green algae contain both simple unicellular members comparable in organisation with any other unicellular protist, but both of these groups as well as the brown algae also contain large thalloid members. The cell differentiation of multicellular organisms is very limited in some filamentous algae and fungi, but becomes complex in higher plants and very complex in metazoa, involving sophisti-cated developmental processes.

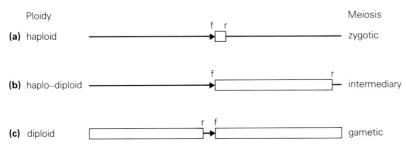

Figure 2.16 The three life cycle patterns of sexually reproducing eukaryotes: f, fertilisation; r, reduction division.

The active feeding cells of some protists are haploid, and undergo repeated division as haploid cells before some of the cells mature as gametes and fuse in pairs at syngamy (fertilisation). The diploid nucleus of the zygote frequently undergoes meiosis in its first divisions (Fig. 2.16a). Such zygotic meiosis in a life cycle that is predominantly haploid is found in dinoflagellates, sporozoa, some green and red algae and some fungi, among others. In representatives of green, red and brown algae, as well as in foraminiferans and some chytrids, meiosis is delayed, the zygote undergoes

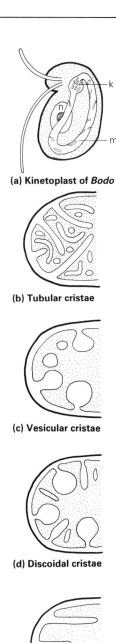

(a) Kinetoplast of *Bodo*

(b) Tubular cristae

(c) Vesicular cristae

(d) Discoidal cristae

(e) Plate-like cristae

Figure 2.17 Mitochondrial features described in the text: k, kinetoplast; m, mitochondrion; n, nucleus.

mitotic divisions before an intermediary meiosis (Fig. 2.16b) and feeding and growth normally occur in both haploid and diploid phases (as in the green land plants, which exhibit a marked alternation of generations). Prolongation of the diploid phase throughout the period of feeding and growth, with restriction of haploid nuclei to the gametes, is familiar in higher animals (Fig. 2.16c); such gametic meiosis is characteristic of diatoms, ciliates, heliozoans, phycomycetes, some brown algae and some flagellate groups. In several protist groups it has not yet been possible to confirm the occurrence of sexual reproduction, or, if this is known to occur, the time of meiosis is uncertain; in these groups, e.g. euglenids, trypanosomes, chrysophytes, cryptophytes and some amoeboid groups, it is not clear whether the organisms are haploid or diploid. *Meiosis appears to be similar in all eukaryotes, except that in a few flagellates it occurs in only one division, without crossing over;* there is no evidence that this indicates a primitive condition, although some of the organisms concerned belong to the Parabasalia.

The essential event in sexual processes is the fusion of two haploid nuclei. This usually, but not invariably, follows the fusion of two haploid cells. When two active haploid cells take part in syngamy they may be structurally identical (**isogametes**), and may be both flagellate (some green algae, some chytrids, some flagellates), or both amoeboid (some heliozoa, some sporozoa and some foraminiferans). When dissimilar gametes (**anisogametes**) fuse, they may both be flagellate, or occasionally both amoeboid. The larger (female) cell is usually non-motile and is called an egg cell and the smaller (male) sperm cell is usually flagellate or rarely amoeboid; fusion of such gametes is referred to as **oogamy** and occurs in brown and red algae, and some members of diatom, green algal, sporozoan, foraminiferan and chytrid groups as well as in higher animals and higher plants. In true fungi and some green algae a process of conjugation occurs when two organisms (usually filamentous) meet and after formation of a connection between the cells cytoplasmic fusion takes place; in the algae and some fungi, nuclear fusion and meiosis follow, but in other fungi the two haploid nuclei may undergo repeated mitosis (in dikaryotic cells) before a delayed fusion.

2.9.3 MITOCHONDRIA, CHLOROPLASTS AND OTHER MEMBRANOUS COMPONENTS OF CELLS

Mitochondria are absent from three phyla of protists, Parabasalia, metamonad flagellates and microspora, most or all of whose members are parasitic, as well as from a number of other anaerobic species or specific cell types. The ancestry of some of these (e.g. the Parabasalia, which have supposedly primitive nuclei) may involve lines that never possessed mitochondrial symbionts, while mitochondria may have been permanently lost by others. In some cases, e.g. the amoeba *Pelomyxa palustris*, many of whose features appear primitive, symbiotic bacteria may perform some of the roles of mitochondria.

Other eukaryote cells contain one or many mitochondria. Although there is only a single mitochondrion per cell in some protists, in the larger cells it forms a loop, or an extended, even branching, filament. The internal structure of these mitochondria varies in two notable ways: the amount of DNA they contain and the form of the cristae (Fig. 2.17a-e). The single mitochondrion of trypanosomes and related flagellates contains one or several large fibrillar bodies, the DNA-rich **kinetoplasts**, which give this group the name

kinetoplast flagellates. Different shapes of cristae characterise different types of cell in the mammalian body; for example the flattened, plate-like cristae of hepatocytes and the tubular cristae of adrenal endocrine cells. The significance of the different shapes is not known.

All chloroplasts are characterised by the presence of two outer membranes surrounding a compartment containing thylakoids and organellar matrix (stroma) which includes circular DNA, ribosomes and sometimes food storage deposits. The basic structure and functioning of chloroplasts are described in detail in another volume in this series. *The arrangement and pigmentation of the thylakoids and the nature and site of the storage substance are variable, and in some groups there are more than two enclosing membranes;* these features give important clues to relationships amongst protists.

The thylakoids of red algal chloroplasts are like those of cyanobacteria; they are separate, with **phycobilisomes** on their outer surfaces and with only chlorophyll a incorporated in their membranes (Fig. 2.18a). In chloroplasts of euglenids, green algae and higher plants the thylakoids are like those of chloroxybacteria in being grouped into stacks, lacking phycobilisomes and containing chlorophylls a and b in their membranes (Figs 2.18 b & e). Phycobilisomes are present within the cisternae of the paired thylakoids of cryptophytes, whose thylakoids membranes contain chlorophylls a and c_2 (Fig. 2.18c). Chlorophylls a and c_2 are also the only chlorophylls in dinoflagellate thylakoids, which normally occur in threes, while in other large photosynthetic groups (diatoms, brown algae and chrysophytes), as well as some lesser ones, the thylakoids also occur in threes but contain chlorophylls a, c_1 and c_2 (Fig. 2.18a).

Photosynthetic products may be stored as starch within the chloroplast, as in green algae and higher plants, or exported to the surrounding cytoplasm and stored there in all other groups. Polysaccharides are stored in the cytoplasm in red algae, dinoflagellates and cryptophytes as starches (α-(1,4)-glucans), or as β-(1,3)-glucan in brown algae (laminarin), euglenids (paramylon), chrysophytes, diatoms and dinoflagellates (chrysolaminarin,

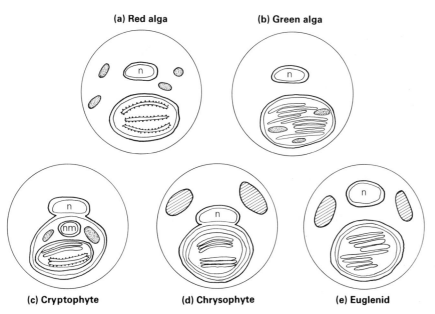

(a) Red alga   (b) Green alga

(c) Cryptophyte   (d) Chrysophyte   (e) Euglenid

Figure 2.18 Diagrams of chloroplasts, their thylakoid arrangement and enclosing membranes in various groups of eukaryotes described in the text. The shaded bodies are food storage structures, starches are shown black and β–(1,3)-glucans are cross-hatched. n, nucleus; nm, nucleomorph.

leucosin). Lipids are also stored as droplets in the cytoplasm of dinoflagellates, diatoms and chrysophytes.

The chloroplasts of red and green algae and metaphytes are surrounded by only two membranes; assuming that the chloroplasts have a symbiotic origin, the membranes are derived from an inner symbiont membrane and an outer host vacuole membrane, and the symbiont in the red alga is cyanobacterial, while that in the green algae and higher plants is chloroxybacterial (Figs 2.18a & b). *No prokaryotes have yet been found that contain chlorophyll c*, but it seems likely that such autotrophic prokaryotes (perhaps several types with different pigment combinations) could have become symbionts in early eukaryotes ancestral to those protist groups that contain chlorophyll c. In cryptophytes the chloroplasts lie in a cytoplasmic compartment that also contains a nuclear remnant (**nucleomorph**) and is separated from the cryptophyte cytoplasm by two further membranes (Fig. 2.18c); this is explained by assuming that the ancestral cryptophyte acquired a eukaryotic symbiont with the unique thylakoid pattern found in this group, and that the outer pair of membranes are derived from the cell membrane of the eukaryote symbiont and the cryptophyte vacuole. Other examples where two pairs of membranes surround the chloroplast (Fig. 2.18d) are also assumed to be derived from symbioses with eukaryotes, and where three membranes are present (Fig. 2.18e), fusion of the two middle membranes is suspected.

The distribution and functions of microbodies in different eukaryotes are incompletely known, but deserve some comment. In mammalian cells microbodies were named **peroxysomes** since they function to oxidise a variety of substrates producing hydrogen peroxide, which is reduced to water under the action of catalase. In the cells of higher plants microbodies were named **glyoxysomes** since their main function is the formation of carbohydrate from fat by the glyoxylate cycle. In the ciliate *Tetrahymena* and in several yeasts (ascomycete fungi) the peroxysomes contain enzymes of the Glyoxylate Cycle. Other microbodies contain quite different collections of enzymes and probably have a different ancestry; for example, the glycosomes of some trypanosomid flagellates contain a collection of glycolytic enzymes. Parabasalian flagellates (which lack mitochondria) contain microbodies which perform a primitive respiratory role by oxidising pyruvate to acetate and CO_2 in a reaction coupled to ATP synthesis; where oxygen is available electrons are transferred via ferredoxin to oxygen and water is formed, but under the anaerobic conditions more usual for these organisms the electrons are transferred to protons, forming hydrogen – these microbodies have therefore been called **hydrogenosomes**.

Vacuoles and lysosomes associated with phagocytic food uptake abound in many heterotrophic cells, but are rarer where only autophagic digestive activity is required. Golgi systems are present in all groups (except perhaps some flagellates), though the number of component cisternae per stack is variable. In cells engaged in active synthesis of secreted components the Golgi stacks are usually more developed and there are more secretory vesicles. Specialised secretory structures include the various types of **extrusomes** produced by diverse protists, and generally serving functions in defence or food capture, e.g. the trichocysts of ciliates; presumably coelenterate nematocysts are more complex secretory structures of the same class.

The main part of the interior of cells of higher plants is occupied by a large vacuole enclosed in a special membrane called the **tonoplast**. Such vacuoles may perform storage functions, retaining, for example, ions or organic substances that cannot be satisfactorily stored in the cytoplasm; *they therefore serve in some ways like intercellular fluids of metazoans.* Large vacuoles of this type are also seen in some green algae and diatoms.

2.10   Cellular evolution amongst eukaryotes

In attempting to trace the development and relationships of eukaryote groups whose cellular features have been outlined in the previous section, it is necessary to consider which features may be primitive and which secondary. For example, it seems clear that the possession of chloroplasts is a *secondary* feature, resulting from different symbiotic events in different sections of the eukaryote adaptive radiation. Photosynthetic members occur in a number of the evolutionary lines, and many of these also contain heterotrophs. There was no primary bifurcation of protists into animals and plants, *but a multiple branching to form many heterotrophic groups, members of several of which gained photosynthetic symbionts and became plants.* In groups which contain both autotrophs and heterotrophs, the heterotrophic lines may never have possessed chloroplasts, or may have lost them; only in some instances do traces of the symbionts remain. Individuals with *plant features may be much more closely related to organisms with animal features than to any other types of plants.* The clear separation between metaphytes with photosynthetic autotrophy, fungi with absorptive heterotrophy and metazoans with phagotrophic heterotrophy does not have phylogenetic relevance among the protists, where, for example, all three nutritional patterns occur amongst euglenid flagellates or dinoflagellates.

Among other indicators of relationships that may be more primitive are features of the nucleus, the cell surface and flagellation. Since there is a tendency in several groups for more primitive members to have closed mitosis and more advanced members to have open mitosis, *it is suggested that closed mitosis is a primitive feature.* It is perhaps more difficult to decide whether internal or external spindles are primitive, but the facts that tubulin necessarily originates in the cytoplasm and groups with external spindles have chromosomal features that are thought to be primitive suggest that external spindles may have occurred earlier, or at least developed in parallel with internal spindles. Similarly, it is assumed that there has been a trend from naked surfaces towards protected ones, either by the development of cell walls in autotrophs and saprotrophs or of pellicles and cytostomes in phagotrophs. Relationships suggested by these features are indicated in Figure 2.19.

Although the possession of a feature may be a more significant indicator than absence, the absence of mitochondria, coupled with apparently primitive nuclear and unspecialised surface features, suggests that parabasalian flagellates may have been an early branch from the eukaryote stock, *possibly originating before mitochondrial symbiosis.* Dinoflagellates, which also have closed mitosis and external spindles, have a relatively complex pellicle which shares several features with the pellicle of ciliates and sporozoans, as well as sharing flagellar and some pellicle features with euglenids and kinetoplastids, all of which have closed mitosis with internal spindles; perhaps this section of ancestral flagellates was developing these pellicular and flagellar features

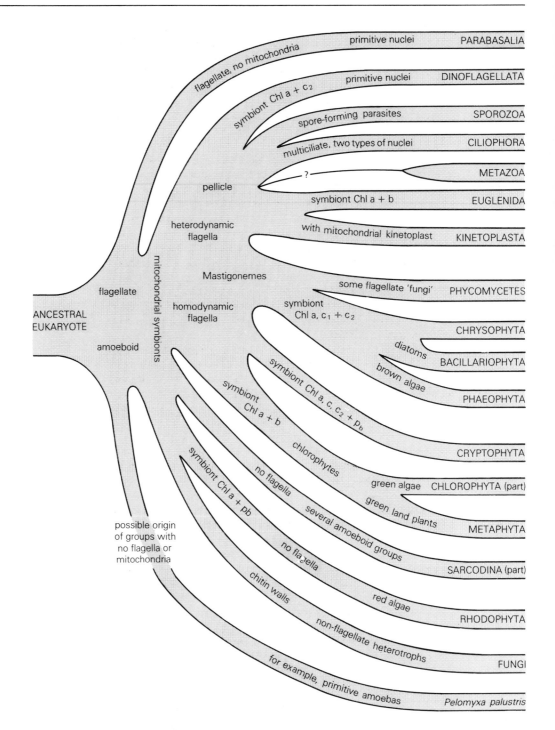

Figure 2.19 A phylogenetic tree of many of the more important eukaryote groups. Many of the suggested links require more evidence to substantiate them. Groups in which all (or nearly all) members are photoautotrophs are cross-hatched; those where only some members are photoautotrophs are shaded.

while the organisation of the spindle was still rather plastic. Flagellate groups of the chrysophyte and chlorophyte lines appear to have had simpler surfaces, more limited phagotrophic ability and smaller body size; they developed different flagellation. The chrysophyte stock gained a symbiont with chlorophylls c_1 and c_2 and gave rise to diatoms, brown algae, phycomycetes and others, including at least some heliozoan amoebae, while the chlorophyte stock acquired a symbiont with chlorophyll b and gave rise to various lines of green algae and land plants.

If the evolutionary advantage that provided an ecological niche for the first eukaryotes was phagocytosis, the presence of mitochondria in some cells and of mitochondria and chloroplasts in others could have provided sufficient energetic advantages for some of these organisms to cease phagocytosis and develop cell walls to live as saprotrophs (fungi) or photoautotrophs (red and brown algae, diatoms and most chlorophytes). In some cases the total absence of flagella may be a primitive feature, and the fungi and red algae, some amoebas and smaller groups of parasitic protozoa, and perhaps one group of green algae, all of which lack flagella, could have originated from an amoeboid stock that never had flagella, other non-flagellate groups perhaps having been derived from different flagellate lines. Other amoeboid groups have flagellate stages, as gametes or migratory phases, or sometimes retain flagella for food collection or locomotion; *these also must have diverged from other protists after flagella were developed.*

The diversification of eukaryote organisation in various directions and the acquisition of mitochondrial and chloroplast symbionts as suggested here could have given rise to the range of cellular patterns characteristic of protists and fungi. The green land plants are assumed to have been derived from a branch of the green algae and show *progressive diversification of cell form in larger plants* to cope with such functions as support, protection and the transport and control of water, nutrients and assimilates. The origins of higher animals are less clear, and there could be *two, three or more lines of origin* from different flagellated protists, leading perhaps to sponges (from collar flagellates) and to a planuloid form, or separately to flatworms and to cnidarians. However they arose, the Metazoa have developed a wealth of diversity in their cellular features, exploiting such capacities as contraction, conduction, sensory reception, secretion, etc., in extreme ways allowed for by the functional interdependence developed within the multicellular body.

All the cells of a multicellular organism normally contain the same genes, but different genes are transcribed in different cells to produce different proteins performing different functions. However, each cell generates its own ATP to energise its activities, generally using its own mitochondria. Some features, like the nucleus, protein synthesis machinery, mitochondria, membranous and fibrous components of cytoplasm are common to every cell in the body, but the emphasis on particular forms of synthesis is varied in different cells to allow the performance of specialised functions.

References and further reading

BOOKS

Alberts, B., D. Bray, J. Lewis, M. Raff, K. Roberts and J. D. Watson 1983. *Molecular biology of the cell.* New York: Garland.
Barnes, R. S. K. 1984. *A synoptic classification of living organisms.* Oxford: Blackwell.
Margulis, L. 1981. *Symbiosis in cell evolution.* New York: W. H. Freeman.

SELECTED ARTICLES

Cavalier-Smith, T. 1975. The origin of nuclei and of eukaryotic cells. *Nature* **256**, 463.
Dickerson, R. E. 1978. Chemical evolution and the origin of life. *Scient. Am.* **239** (Mar.), 30.
Dickerson, R. E. 1980. Cytochrome c and the evolution of energy metabolism. *Scient. Am.* **242** (Mar.), 98.
Woese, C. R. 1981. Archaebacteria. *Scient. Am.* **244** (June) 94.

Cell Biology

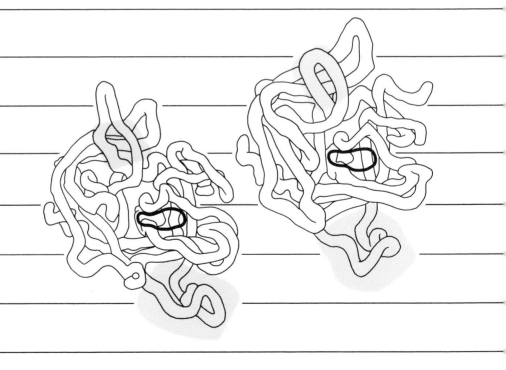

3

Enzymes

C. J. Wynn

*MODERN
VIEWS IN
BIOLOGY*

3.1 Role of enzymes

To the biochemist, the cell may be regarded as a highly sophisticated factory. It derives its raw materials either from some circulatory system or directly from its external environment, depending on the degree of complexity of organisation of the organism, and produces a great variety of end products from these basically simple materials. In any cell there are many thousands of different compounds present, ranging from simple inorganic ions to the extremely large and complex molecules of DNA. As well as this production capacity, the cell possesses the ability to divide and reproduce itself, to replace and repair its basic structure continuously, to communicate and act in concert with other cells and to maintain the concentrations of the many compounds it contains within relatively narrow limits. When to this diversity of functions is added the ability to respond extremely rapidly to a variety of internal and external stimuli, the tremendous complexity of the cell at the chemical level can be appreciated.

While the information for the performance of these many functions is inherited by the cell in the form of DNA (see §5.1.2) each cell makes its own machinery to carry them out. *This machinery is the enzymes of the cell* and the aim of this chapter is to show how enzymes, by virtue of their structure, are able to carry out the tasks allotted to them in a highly specific, very rapid and closely controlled manner.

3.1.1 PROTEINS AS CATALYSTS

Some of the reactions that occur within the cell are easily performed in the test-tube. For instance, if carbon dioxide is bubbled through water, some of it dissolves and, in doing so, it reacts with the water to form carbonic acid thus:

$$H_2O + CO_2 \rightarrow H_2CO_3$$

Although this reaction is spontaneous and occurs relatively rapidly, in the red blood cell it is catalysed by an enzyme, carbonic anhydrase, which adds the property of *controllability* to this reaction.

On the other hand, some cellular reactions occur only very slowly *in vitro*. The breakdown of hydrogen peroxide to oxygen and water

$$2H_2O_2 \rightarrow 2H_2O + O_2$$

occurs only slowly at room temperature but can be speeded up by the addition of either H^+ ions or Fe, i.e. **inorganic catalysts**. The enzyme catalase, which is present in many cells (see §2.9.3) and whose function is to destroy any hydrogen peroxide produced in the cell before it can oxidise any neighbouring susceptible molecules, is able to catalyse this reaction so that its rate is 10^4 times greater than the H^+-catalysed reaction, which in turn is 10^4 times faster than the uncatalysed reaction. In addition, once peroxide breakdown is initiated in the test-tube by the addition of acid, the reaction proceeds to completion and cannot be readily stopped. However, the action of the enzyme can be controlled by alteration of various parameters, and hence the breakdown of the peroxide can be made responsive to the needs of the cell.

Finally, other cellular reactions occur at an immeasurably slow rate *in vitro*. Solutions of DNA and RNA may be stored under aseptic conditions in a refrigerator for several months without appreciable deterioration and hydrolysis and yet may be rapidly and specifically degraded by the nuclease group of enzymes.

All enzymes are proteins and, although in many cases the protein moiety may be modified by the addition of carbohydrate to form glycoproteins or lipid to form lipoprotein, the catalytic power of the enzyme resides in the protein part. As we shall discuss later, *the carbohydrate or lipid modification serves in general to direct the šbcellular localisation of the enzyme or to assist its insertion into specific structures, such as membranes,* but rarely has any significant influence on the catalytic process. To understand the nature of this catalytic mechanism, we first examine in some detail the molecular structure of proteins.

3.2 Levels of protein structure

The protein as synthesised on the ribosome (see §6.8) is a linear sequence of amino acids, polymerised by the elimination of water between successive amino acids to form the peptide bond, and existing as a randomly coiled chain without specific shape and possessing no biological, i.e. catalytic, activity. Within seconds of synthesis being completed, *the protein folds into a specific three-dimensional form,* which is the same for all molecules of the same protein and which now is capable of catalysis in many cases. What are the factors responsible for this highly specific folding?

This is part of the more general question, do molecules exist *randomly* or are there forces that allow interaction in a *controlled manner* between different molecules of the *same* compound, between molecules of *different* compounds and even between *different parts* of the same molecule, if it is sufficiently large? The field of organic chemistry depends on the ability of carbon atoms to form stable covalent bonds with other carbon atoms and with a variety of other atoms including nitrogen, oxygen and hydrogen. On the other hand, sodium chloride, even in the crystal, exists in the form of Na^+ and Cl^- ions held together by the interaction of unlike charges, the lattice energy. In solution, the lattice energy is overcome and the molecule ionises, the natural tendency of the unlike charges to associate together being counteracted by the ability of both ions to attract to themselves molecules of water, which are polar and present in large excess in such a solution. In this way, a variety of forces compete to determine the structure of even as simple a thing as a solution of sodium chloride.

There are four non-covalent interactions that are important in the production of the specific three-dimensional configuration of a protein.

(a) **Ionic interactions** which occur between either fully charged groups (such as those in the NaCl crystal) or between partially charged groups (such as the interaction of a polar water molecule with a Na^+ ion in solution).

(b) **Hydrogen bonds** where a hydrogen atom is shared between two other atoms (both electronegative, e.g. oxygen and nitrogen). A good example of this type of interaction is the structure of water itself where the electrons in the covalent O-H bond are associated more with the oxygen nucleus than the H nucleus producing a polarity in the bond, usually represented by partial charges. These partial charges on neighbouring molecules interact giving liquid water a three-dimensional structure thus:

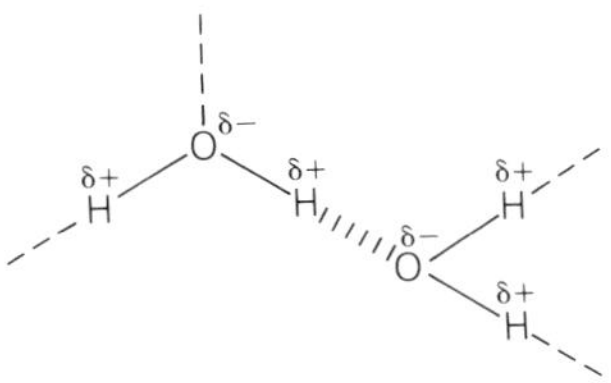

All life depends on water having this structure, for, in the absence of hydrogen bonding, water would be gaseous at room temperature.

(c) **Hydrophobic** interactions arise from the tendency of non-polar groups to interact together to minimise the proximity of polar groups. Emulsions of oils in water separate easily on standing with the oil, forming large globules on the surface of the water. In this way there is minimal interference with the hydrogen bonding in water, and oil molecules are satisfied by the formation of a local hydrophobic environment. This type of behaviour is shown by molecules containing hydrocarbon chains, which are essentially non-polar.

(d) **Van der Waals'** interactions occur when any two atoms approach each other very closely. The fluctuating electrical charges set up between the nuclei and the electron 'clouds' create a weak ionic interaction. Although individually these van der Waals' interactions are weak, they become highly significant when the surfaces of two macromolecules fit closely together.

3.2.1 PRIMARY STRUCTURE

The primary structure of a protein is its *linear sequence of amino acids*. Of the 20 commonly occurring amino acids, 19 may be represented by the general formula

$$R$$
$$|$$
$$NH_2—CH—COOH$$

the exception being proline, where the amino group forms part of a ring structure. Because these α-amino and carboxyl groups are involved in peptide bond formation, there will be only one of each free per protein chain at the so-called **N-terminal** and **C-terminal** ends respectively. The tremendous diversity of proteins is explained by the variation in the nature of the R sidechain (Fig. 3.1).

In certain amino acids R is either a hydrogen atom (glycine, abbreviated Gly) or a hydrophobic aliphatic (leucine, Leu) or aromatic (phenylalanine, Phe) hydrocarbon. In other cases, R contains either an extra carboxyl group or an extra amino group or its equivalent. Glutamic acid (Glu) and aspartic acid (Asp) each have an extra carboxyl group and, as this group is ionised at physiological pH, these particular amino acid residues in a protein chain will

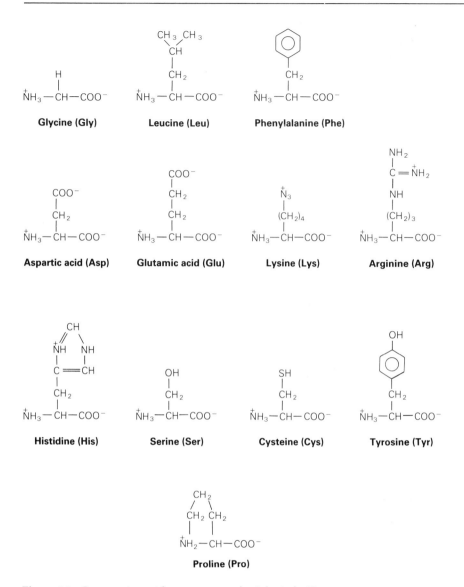

Figure 3.1 Some amino acid structures at physiological pH.

be negatively charged (hence they are often called glutamate and aspartate respectively). Lysine (Lys) and arginine (Arg) both contain an additional amino group or equivalent structure and, as this is protonated at physiological pH, are positively charged. Histidine (His) also contains a protonatable N, this protonation occurring over the physiological range of pH, and hence this amino acid residue may or may not be charged. Other amino acids such as serine (Ser) and tyrosine (Tyr) have hydroxyl groups in their sidechain and this permits the ready formation of hydrogen bonds. Of particular importance is the amino acid cysteine (Cys) which possesses a thiol (SH) group. In the oxidising environment of the endoplasmic reticulum, two such cysteine residues in adjacent protein chains or in different parts of the same chain join together with the elimination of two hydrogen atoms to form a disulphide bridge.

$$\text{Cys—S—S—Cys}$$

This bridge helps to stabilise and protect the protein structure and is *commonly found in those proteins that are secreted from the cell and function in the less controlled extracellular environment.*

That the primary structure, and only the primary structure, of a protein decides its three-dimensional form was shown in a series of elegant experiments by Anfinsen in the early 1960s. He reduced the disulphide bridges in the enzyme ribonuclease from pancreatic juice using β-mercaptoethanol and unfolded the molecule by treatment with high concentrations of urea, which minimises many intramolecular interactions. The unfolded, denatured protein was devoid of catalytic activity but this

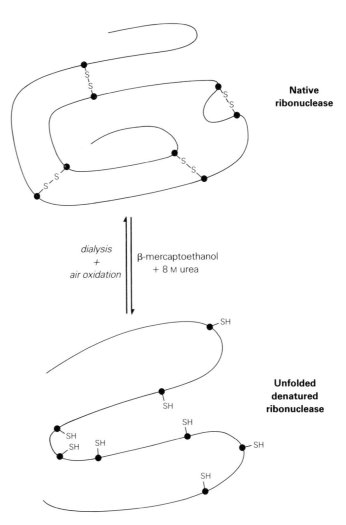

Figure 3.2 The denaturation and renaturation of ribonuclease, emphasising the role of disulphide links in the stabilisation of protein structure.

activity was restored when the protein solution was dialysed to remove the small urea and β-mercaptoethanol molecules and allowed to stand in air for a short while (Fig. 3.2). Thus the information for the correct folding of the protein was present in the random unfolded protein chain and, given the correct conditions, the protein could refold without the aid of any external factors. Anfinsen concluded that *the native form of ribonuclease was the thermodynamically most stable structure.*

Folding must occur by alteration of angles between atoms in the peptide backbone. If the angle between adjacent amino acid residues has only two permissible values, the number of structures that a protein containing 200 amino acid residues can adopt is 2^{200}. If the time to flip from one structure to another is 10^{-12} s, the time taken for the protein to fold would be $10^{60} \times 10^{-12}$ s or approximately 10^{40} years. Yet a protein can fold in a test-tube in minutes. How is this achieved without any external influence?

3.2.2 SECONDARY STRUCTURE

In 1951 Pauling and Corey proposed that certain portions of protein molecules could fold into regularly repeating structures and they identified two such probable structures: the α-**helix** and the β-**pleated sheet**. Their postulates were based on molecular models of polypeptides and they adopted the criteria that for a stable structure to be formed it should contain the maximum number of hydrogen bonds between oxygen and nitrogen atoms in the peptide backbone and should allow the oxygen, hydrogen and nitrogen atoms of the hydrogen bonds to lie in a straight line. These theoretical secondary structures were not validated by experiment until more than six years later when Kendrew and Perutz in their X-ray crystallographic studies of protein crystals were able to demonstrate their existence (see §1.9). We now know that substantial portions of many proteins form helices and pleated sheets (Fig. 3.3). Certain amino acids such as glutamic acid and leucine favour the formation of helix, while proline with its unusual structure is a helix breaker. On the other hand, amino acids with a bulky sidechain such as tyrosine are more readily accommodated in a pleated sheet structure.

As the protein starts to fold, short sections of secondary structure will be formed transiently and, because of their inherent stability due to H-bonding, will tend to persist and form a nucleus for the growth of further helix or pleated sheet. Interactions between neighbouring helices and/or pleated sheets then continue with consequent stabilisation. Hence this theory postulates that the *formation of stable secondary structures is the initial stimulus to protein folding.* These transient intermediates in protein folding have not yet been detected but the theory is attractive in explaining other demonstrable facets of protein structure.

The development of extremely powerful computers in recent years has allowed the simulation of protein folding to be studied. It has been found that if, in the initial stages of folding, the computer is instructed to search for short sequences of amino acids capable of forming secondary structure, then its prediction of the final three-dimensional configuration is a good approximation to that found by X-ray crystallography. Preliminary experiments on the prediction of the structure of unknown proteins by computer folding have been validated by subsequent experiment and this technique is likely to provide a very powerful tool in the study of protein structure and the interactions of such structures with other macromolecules in the cell.

Figure 3.3 (a) The α-helix showing the stabilisation of the structure by intra-chain hydrogen bonding. (b) Antiparallel β-pleated sheet. For clarity, the C_α and R substituents are omitted in both diagrams.

(a)

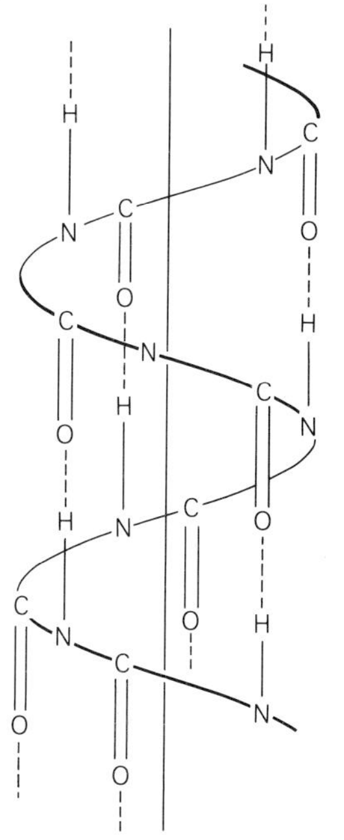

(b)

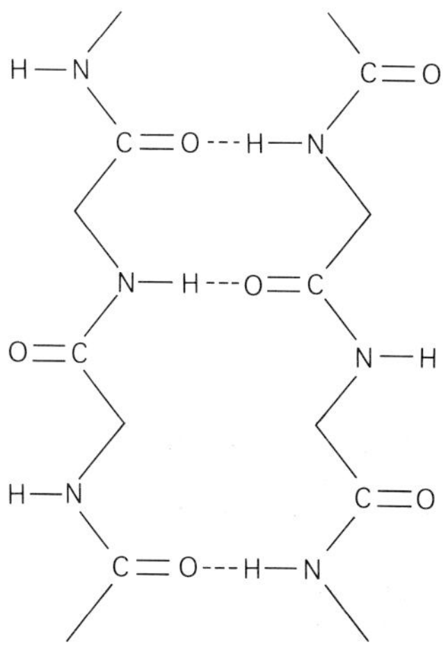

In addition to its role in the initiation of folding, secondary structure seems also to have biological significance. As more information has been obtained about the three-dimensional structure of protein molecules, it has become apparent that *many proteins with similar functions possess very similar structures.* The most obvious example is the near superimposability of the α and β chains of haemoglobin and the myoglobin molecule. The oxygen-binding capacity of each of these molecules is dependent on the presence of a non-protein molecule, haem, which is located in a hydrophobic pocket of the globin structure, and, being non-covalently bound, may be removed from the molecule by relatively mild procedures such as lowering the pH. Such a molecule is termed a **prosthetic group** and there are many other examples of such non-protein, loosely bound molecules. If the haem is removed from a haemoglobin molecule it loses its ability to bind oxygen showing that the *hydrophobic environment is absolutely essential for biological activity.* The presence of eight major helical segments in all oxygen-carrying proteins suggests that such a secondary structure has evolved as the perfect hydrophobic environment for the haem group. Similarity of secondary structure does not necessarily imply identity of primary structure. In fact, only 23 residues in β-globin chains are common to the 141 residues in α-globin chains. *Thus quite different amino acid sequences can specify very similar three-dimensional structures.*

Another example of striking similarity in secondary structure is shown by the family of enzymes known as the serine proteinases. This family of protein-hydrolysing enzymes includes the digestive enzymes chymotrypsin, trypsin

Figure 3.4 Comparison of the structures of elastase and chymotrypsin: (a) Three-dimensional structure showing dimensions and regions of anti-parallel pleated sheet; (b) Primary structure aligned for maximum homology. The homologous areas are shaded.

(a)

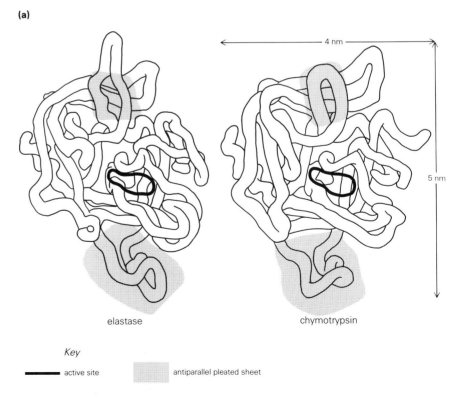

(b)

C = chymotrypsin   E = elastase

		Cys	Gly	Val	Pro	Ala	Ile	Gln	Pro	
C	Val	Leu	Ile	Val	Asn	Gly	Glu	Glu	Ala	
E			Ile	Val	Gly	Gly	Thr	Glu	Ala	
C	Val	Pro	Gly	Ser	Trp	Pro	Trp	Glu	Val	
E	Gln	Arg	Asn	Ser	Trp	Pro	Ser	Glu	Ile	
C	Ser	Leu	Glu	Asp	Lys	Thr		Ser	Ser	
E	Ser	Leu	Glu	Tyr	Arg	Ser	Gly	Ser	Ser	
C	Gly	Phe	His	Phe	Cys	Gly	Gly	Ser	Leu	
E	Trp	Ala	His	Thr	Cys	Gly	Gly	Thr	Leu	
C	Ile	Asn	Glu	Asn	Trp	Val	Val	Thr	Ala	
E	Ile	Arg	Gln	Asn	Trp	Val	Met	Thr	Ala	
C	Ala	His	Cys	Gly	Val	Thr		Thr	Ser	
E	Ala	His	Cys	Val	Asp	Arg	Glu	Leu	Thr	
C	Asp	Val	Val	Val	Ala	Cys	Glu	Phe	Asp	
E	Phe	Arg	Val	Val	Val	Cys	Glu	His	Asn	
C	Glu	Gly	Ser	Ser	Ser	Glu	Ile	Ile	Gln	
E	Leu	Asn	Gln	Asn	Asn	Gly	Thr	Glu	Gln	
C	Lys	Ile	Lys	Ile	Ala	Lys	Phe	Phe	Lys	
E	Tyr	Val	Gly	Val	Gln	Lys	Ile	Val	Val	
C	Asn	Ser	Lys	Tyr	Asn	Ser	Leu	Thr	Ile	
E	His	Val	Tyr	Trp	Asn	Thr	Asp	Asp	Val	
C			Asn	Asn	Asp	Ile	Thr	Leu		
E	Ala	Ala	Gly	Tyr	Asp	Ile	Ala	Leu		
C	Leu	Lys	Leu	Ala	Thr					
E	Leu	Arg	Leu	Ala	Gln					
C	Ala	Ala	Ser	Phe	Ser	Gln	Thr	Val	Ser	
E	Ser	Val	Thr	Leu	Asn	Ser	Tyr	Val	Gln	
C	Ala	Val	Cys	Leu	Pro	Ser	Ala	Ser	Asp	
E	Leu	Gly	Val	Leu	Pro	Arg	Ala	Gly	Thr	
C	Asp	Phe	Ala	Ala	Gly	Thr	Thr	Cys	Val	
E	Ile	Ile	Ala	Asn	Asn	Ser	Pro	Cys	Tyr	
C	Thr	Thr	Gly	Trp	Gly	Leu	Thr	Arg	Tyr	Asn
E	Ile	Thr	Gly	Trp	Gly	Leu	Thr	Arg	Thr	Asn
C	Ala	Ala	Asn	Thr	Pro	Asp	Arg	Leu	Gln	Gln
E		Gly	Gln	Leu	Ala	Gln	Thr	Leu	Gln	Gln
C	Ala	Ser	Leu	Pro	Leu	Leu	Ser	Asn	Thr	Asn
E	Ala	Tyr	Leu	Pro	Thr	Val	Asp	Tyr	Ala	Ile
C	Cys	Lys	Lys		Ser	Tyr	Trp	Gly	Thr	Lys
E	Cys	Ser	Ser	Ser	Ser	Tyr	Trp	Gly	Ser	Thr
C	Val	Lys	Asp	Ala	Met	Ile	Cys	Ala	Gly	Ala
E	Val	Lys	Asn	Ser	Met	Val	Cys	Ala	Gly	Gly
C		Ser	Gly	Val	Ser	Ser	Cys	Met	Gly	
E	Asp	Gly	Val	Arg	Ser	Gly	Cys	Gln	Gly	
C	Asp	Ser	Gly	Pro	Leu	Val	Cys	Lys	Lys	
E	Asp	Ser	Gly	Pro	Leu	His	Cys	Leu	Tyr	
C	Asn	Gly	Ala	Trp	Thr	Leu	Val	Gly	Ile	
E	Asn	Gly	Gln	Tyr	Ala	Val	His	Gly	Val	
C	Val	Ser	Trp	Gly	Ser	Ser		Thr	Cys	
E	Thr	Ser	Phe	Val	Ser	Arg	Leu	Gly	Cys	
C	Ser	Thr	Ser		Thr	Pro	Gly	Val		
E	Asn	Val	Thr	Arg	Lys	Pro	Thr	Val		
C	Tyr	Ala	Arg	Val	Thr	Ala	Leu	Val	Asn	
E	Phe	Thr	Arg	Val	Ser	Ala	Leu	Ile	Ser	
C	Trp	Val	Gln	Gln	Thr	Leu	Ala	Ala	Asn	
E	Trp	Ile	Asn	Asn	Val	Ile	Ala	Ser	Asn	

and elastase (a pancreatic endopeptidase), together with thrombin and other enzymes that control the clotting of blood. X-ray crystallography (see §1.9) shows that if two of these enzymes are compared, the details of the secondary structure are nearly identical in spite of only about 40 per cent of the amino acid residues having a common sequence (Figs 3.4a & b). In the course of evolution, genes can be duplicated and modified. If a protein with a stable three-dimensional structure is produced, this structure can often be incorporated in a variety of other proteins and by suitable modification of parts of the amino acid sequence can be made to catalyse a different, although similar, reaction. In this way it is believed *that families of proteins may have evolved from a single ancestral gene* (see §5.4 and 7.4).

A specific secondary structure usually with a specific function is often referred to as a '**domain**'. Many proteins have a '**multidomain**' structure where each domain has retained its specific function. For example, there is a family of enzymes called the dehydrogenases whose function is to catalyse the oxidation of a variety of intermediary metabolites. These have a common hydrogen acceptor, nicotinamide adenine dinucleotide (NAD). Such molecules, which serve the same function in a variety of reactions and which are usually regenerated by a subsequent reaction, making them required in only small quantities, are called **coenzymes**. The NAD domain of a dehydrogenase is the site of binding of NAD and each dehydrogenase will have a further domain where the molecule to be oxidised is bound. It seems probable that the evolutionary strategy has been to create amino acid sequences by joining existing ones together. Such a strategy will be *far more efficient than chance mutation of DNA structures to give a new protein sequence.*

3.2.3   TERTIARY STRUCTURE

The spatial orientation of the various sections of secondary structure is called the **tertiary structure** and it is this structure which gives a protein its final three-dimensional form. The predominant force in the folding of a protein into its tertiary structure is the *interaction between hydrophobic groups in the amino acid sidechain.* In order to minimise their interaction with the aqueous environment of the protein in solution, these groups interact to produce a hydrophobic milieu from which water is excluded. In this way, it is found that hydrophobic sidechains of a protein molecule are found in the centre of its three-dimensional structure while polar groups which are readily solvated are on the surface of the molecule. The rest of the folding of the molecule depends on ionic interactions, e.g. between charged lysine and aspartate groups and hydrogen bonding between suitable electronegative atoms in the sidechains. Ultimately each protein folds in a specific secondary and tertiary structure *such that the most thermodynamically stable molecule is achieved.* For naturally occurring proteins, the native three-dimensional configuration is by far the most stable structure that can be formed from that specific amino acid sequence. This stability protects the protein from minor energy fluctuations in the cell and ensures continuity of biological function.

Recent estimates suggest that there may be as many as 10 000 proteins present in any species. However, very many more different sequences of amino acids can theoretically exist. If one considers a peptide chain containing 5 amino acids, it is possible to select each of the amino acids from the 20 commonly occurring, and since the order is important, the total number of pentapeptides that can be made is theoretically $20 \times 20 \times 20 \times 20 \times 20 = 3\ 200\ 000$. Imagine, if you can, the possibilities for proteins which on average contain at least 200 amino acid residues. Modern computer calculations have shown that the vast majority of these sequences would each fold into many different structures with about equal energy and possessing very different chemical properties. They would be inherently unstable and could easily change from one structure to another and hence lose their specific function. The cell is a highly ordered structure that cannot function efficiently with such vacillating catalysts and, if they arise during chance mutation, they are probably lost by natural selection.

3.2.4 QUATERNARY STRUCTURE

The functional unit of many proteins is not a single polypeptide chain but rather a series of chains (**sub-units**) aggregated together to form the active protein (**multimer**). In some proteins all the sub-units are identical and the aggregation may protect the enzyme against too rapid degradation in the cell or may be necessary for the development of full catalytic activity. In other cases, two or three different sub-units may interact to form the functional multimer, and it is often found that while one sub-unit has catalytic activity, another sub-unit will have a regulatory function and will allow the catalytic activity to be modified in response to the changing needs of the cell (e.g. RNA polymerase, §6.3). While multimeric enzymes containing up to 32 sub-units have been described, the most common multimers are dimers, trimers and tetramers.

The formation of multimeric structures provides an alternative strategy for the evolution of enzymes. In the 'multidomain' enzymes described previously, the splicing together of DNA segments to give a single polypeptide chain with at least two well defined stable functional parts provided a method for the elaboration of new and useful enzymes during the evolution of the cell. By the *selection of sub-units which have specific binding capacities and are able to interact with each other non-covalently to form a stable structure, it is possible to evolve another series of enzymes capable of fine control.*

3.3 Enzyme specificity

3.3.1 DEGREE OF SPECIFICITY

Enzymes are specific catalysts – specific in the nature of the reaction catalysed and specific in the actual reactant (**substrate**) molecules. Within any one family of enzymes there is a fairly wide variation in the degree of specificity. Taking as an example the proteolytic enzymes, the bacterial enzyme, subtilisin, cleaves any peptide bond regardless of the amino acids which it joins, while at the other extreme thrombin catalyses the hydrolysis only of peptide bonds between arginine and glycine. Other proteolytic enzymes have intermediate specificity, e.g. trypsin cleaves peptide bonds on the carboxyl side of lysine and arginine. Such specificity differences are mirrored in the biological function of the various proteolytic enzymes. Bacteria need the ability to use any protein as a source of carbon and nitrogen. Thus the ability to digest any protein with which they come in contact is useful. Mammalian intestinal digestion of proteins proceeds using a battery of enzymes (pepsin, trypsin and chymotrypsin) with limited specificities which break down the protein into small fragments usually 5-20 amino acids long. Carboxypeptidases and aminopeptidases are then able to split off one amino acid at a time from either end until all peptide bonds have been broken. To allow efficient digestion and utilisation of all ingested proteins, the wide specificity of this group of enzymes is important. Thrombin, on the other hand, splits a single peptide bond in fibrinogen, releasing peptide fragments and leaving the fibrin monomer which is able to polymerise and form a fibrin clot which arrests haemorrhage from a blood vessel. Thus thrombin is single-minded in its purpose and this is reflected in its *absolute specificity*.

In general, the hydrolases are probably the least specific class of enzymes. Other classes of enzymes, such as those catalysing the transfer of groups between molecules, oxidation and reduction, the joining of two molecules or the interconversion of two isomers, have far narrower specificities, being able to differentiate between chemical isomers such as glucose and galactose and between stereo-isomers such as D and L glucose.

3.3.2   MOLECULAR BASIS OF SPECIFICITY

The molecular basis of specificity ultimately depends on the three-dimensional structure of the enzyme molecule. Enzymic catalysis can be considered as taking place in three stages:

(a) binding of substrate to the enzyme;
(b) catalysis of the reaction depending on the close proximity of the substrate and specific amino acid residues; and
(c) release of products from the enzyme.

Every enzyme has two closely situated sites – the binding and the catalytic sites – which together constitute the **active site** or **active centre** of the enzyme. Since the spatial juxtaposition of the substrate and its enzyme is so precise, even minute alterations in the active centre due to small changes in the three-dimensional structure are often sufficient to cause the enzyme to lose its catalytic activity.

These considerations are well illustrated by comparing the substrate-binding sites of chymotrypsin, which hydrolyses peptide bonds after aromatic sidechains, and trypsin, which hydrolyses peptide bonds after lysine and arginine. In chymotrypsin, there is a hydrophobic 'pocket' into which the bulky, non-polar, aromatic amino acid can fit (Fig. 3.5). In trypsin, a serine residue in this pocket is changed into a polar, negatively charged aspartate residue. This repels non-polar amino acids but readily attracts positively charged residues such as lysine and arginine.

3.3.3   REACTION RATES AND PURIFICATION

The direct measurement of the amount of any enzyme is usually impossible. Some enzymes contain rare and specific groups such as a metallic ion, but only in purified enzyme preparations is it justified to assume that no other component contains such features, and hence to use this feature to estimate the enzyme directly. Commonly, the rate of the catalysed reaction is measured

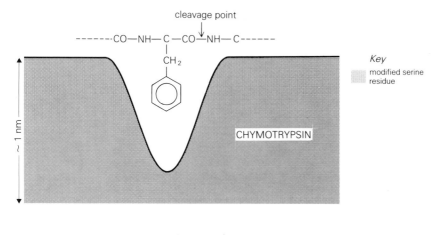

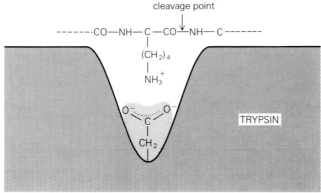

Figure 3.5 The hydrophobic 'binding pocket' of chymotrypsin contrasted with the negatively charged pocket of trypsin. A small segment of peptide is shown bound in each case.

and, after correcting for any uncatalysed reaction rate, the enzyme activity is calculated as a function of the rate of the catalysed reaction. The **International Unit** (IU) of enzyme activity is hence defined as that amount of enzyme that will catalyse the conversion of 1 μmole of substrate to product in 1 min under defined conditions. Thus basically the problem of measuring enzyme activity becomes the problem of measuring reaction rate. Where there is a sharp difference in physical properties between substrate and product, e.g. ultraviolet or visible light absorption, viscosity, fluorescence, physical state or acidity, then continuous measurement of that property as a function of time enables the reaction rate to be measured directly. In other cases it is necessary to allow the reaction to proceed for a short measured time before separating the reactants from products and estimating either chemically. Such indirect measurements rely for their validity on the specificity of the enzyme. If there is more than one enzyme present which can catalyse a certain reaction, and occasionally this can happen, then some other variable must be introduced into the reaction system, such as an inhibitor, to allow the measurement of the separate enzymes.

Specificity is also important in the purification of enzymes. Separation procedures based on size, charge and solubility (see §1.5) can partially purify an enzyme from a crude extract. However, because in the multiplicity of proteins present in any extract there are likely to be quite a few with nearly identical size and charge, these separation procedures are rarely sufficient to allow the enzyme to be purified to homogeneity. To allow complete purification, recourse has to be made to methods involving the specificity of the enzyme. By binding the substrate or some other ligand that is able to recognise and show complementarity to the enzyme, e.g. competitive inhibitor, to an inert support and passing impure enzyme down a column of such material, the enzyme is retarded by specific binding to the ligand while impurities being incapable of specific interaction pass straight through. The enzyme may then be eluted by changing the pH or ionic strength to decrease the binding of the enzyme and its immobilised ligand. Such methods, termed affinity chromatography (see §1.6.4), have now been developed and applied to a large variety of enzymes, in some cases achieving over 2000-fold purification in one step. A variety of inert supports such as glass and dextran beads have been used, and the main technical limitation in the extension of the method to all enzymes is the lack of a suitable ligand or suitable binding technique for derivatisation of the inert support.

3.3.4 THE ENZYME-SUBSTRATE COMPLEX

If we consider a simple reaction such as $A \rightarrow B$, then it can be shown that the reaction goes through a **transition state** that has a higher energy than either A or B. The rate of the forward reaction depends on the temperature and on the difference in free energy between that of A and the transition state. This energy difference is called the **Gibbs free energy of activation** and is given the symbol $\triangle G$ (Fig. 3.6a). Although we talk of the energy of A, the individual molecules of A have a range of energies distributed about a mean, which is the value that we usually assign to the energy of A. At any given temperature, *there will be a small but constant proportion of molecules with sufficiently high energy to surmount the transition state energy barrier and form molecules of B.* In some cases this proportion may be so small that the formation of B is too low to be measured. As the temperature is increased, the mean energy of the molecules increases also and with it the proportion of molecules having sufficient energy to surmount the transition state barrier. Thus the rate of formation of B is increased.

When enzyme and substrate combine together, a new reaction pathway is created and this has a lower $\triangle G$ of activation than the uncatalysed reaction (Fig. 3.6b). At any given temperature the proportion of molecules having sufficient energy to surmount this barrier is greater than that for the uncatalysed reaction. *Enzymes accelerate reactions by decreasing the activation energy barrier.*

Inherent in this argument about the role of an enzyme in reducing the transition state barrier and also in the concepts of specificity of enzyme-catalysed reactions, is the assumption that the enzyme and substrate combine together to form a specific complex.

There is now abundant evidence of the existence of such complexes. The complexes have been demonstrated directly in several cases by electron microscopy and X-ray crystallography. The polymerases responsible for the synthesis of nucleic acids (see §6.3) have been shown to bind to a specific region of the substrate molecule. Enzymes have been crystallised in the cold in the presence of their substrate and the resulting X-ray crystallographic data clearly show substrate molecules specifically orientated at the active centre of the enzyme.

In addition to this direct evidence, there is substantial indirect evidence for the existence of an enzyme-substrate complex. Various physical properties of the enzyme alter in the presence of the substrate. The heat stability of an enzyme is *often far greater in the presence of substrate than in its absence.* This increased stability can be explained by the substrate combining with the enzyme at its active site and hence protecting and maintaining the three-dimensional configuration at a higher temperature than normal. Some enzymes contain a metal ion such as Fe^{3+}, Cu^{2+} or Mo^{4+} as part of their active centres. These metallic ions give rise to absorption in the visible region of the spectrum and the spectra often change in the presence of substrate, again providing indirect evidence of the formation of an enzyme-substrate complex.

In some cases it is possible to isolate the complex. This isolation relies on the fact that while the first step in catalysis, the binding of enzyme and substrate, is often rapid, *the next step, the catalysis of the reaction, is often slow.* If an enzyme-substrate mixture is cooled in ice or something even colder while the complex is formed, the subsequent bond cleavages may not occur to any great extent and the complex may accumulate. In other cases, the formation of the complex may involve covalent bonding of the enzyme and substrate, and although this complex may have only a transient existence, it can sometimes be stabilised by chemical methods such as reduction of unsaturated linkages that are formed. We shall see examples of such complexes when we consider the mechanism of action of such enzymes as phosphoglucoisomerase and aldolase in a later section (see §3.4).

Figure 3.6 Energy changes during the course of the reaction $A \rightarrow B$. (a) Uncatalysed reaction; (b) Enzyme catalysed reaction with lower activation energy.

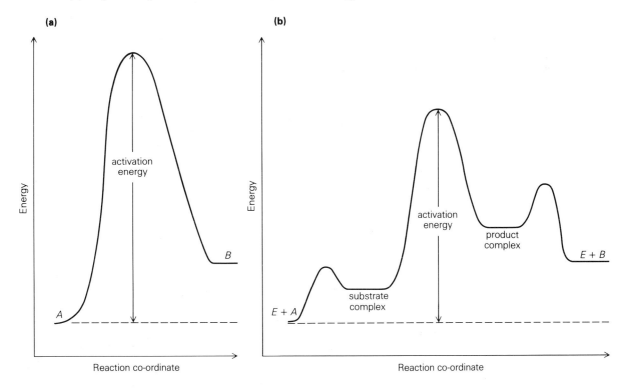

3.3.5 ENZYME KINETICS

In the simplest case where one molecule of substrate, S, is converted to one molecule of product, P, it is possible to write the following reaction mechanism:

$$E + S \underset{k_{-1}}{\overset{k_{+1}}{\rightleftharpoons}} ES \underset{k_{-2}}{\overset{k_{+2}}{\rightleftharpoons}} E + P$$

where each of the two parts of the reaction is shown as reversible and k_{+1} and k_{+2} represent the rate constants of the forward reactions and k_{-1} and k_{-2} represent the rate constants of the reverse reactions. For simplicity the conversion of substrate to product and the release of product from the enzyme have been combined into one step. For many reactions k_{-2} is very small, i.e. the rate of formation of substrate from product, the back reaction, is negligible.

After the initial stages of the reaction, a steady state is set up where the concentration of the enzyme-substrate complex is constant. At this steady state the rate of formation of the complex is equal to the rate of its breakdown, and allowing [E], [S], [ES], [P] to represent the concentrations of enzyme, substrate, complex and products respectively, we can write

$$k_{+1}[E][S] = k_{-1}[ES] + k_{+2}[ES]$$
$$= [ES] (k_{-1} + k_{+2})$$

whence

$$[ES] = [E][S] \quad \frac{k_{+1}}{k_{-1} + k_{+2}}$$

The expression $(k_{-1} + k_{+2})/k_{+1}$ is known as the **Michaelis constant** and is given the symbol K_m. We can now rewrite the steady state equation as

$$[ES] = \frac{[E][S]}{K_m} \qquad (3.1)$$

Enzyme is present both as free enzyme and as enzyme-substrate complex so that

$$[E] = [E_{tot}] - [ES]$$

where E_{tot} is the **total enzyme concentration**.

Substituting this value of E into Equation 3.1 we get

$$[ES] = \frac{([E_{tot}] - [ES])[S]}{K_m}$$

which rearranges to

$$[ES] = [E_{tot}] \left(\frac{[S]}{[S] + K_m} \right)$$

The rate of the measured reaction, the rate of formation of products, is given by

$$v = k_{+2}[ES]$$

$$= k_{+2} [E_{tot}] \frac{[S]}{[S] + K_m} \qquad (3.2)$$

When the substrate concentration is very much greater than K_m the maximum rate of reaction, V, is reached, where

$$V = k_{+2}[E_{tot}]$$

and is independent of substrate concentration. Equation 3.2 now becomes

$$v = V \frac{[S]}{[S] + K_m}$$

which is known as the **Michaelis-Menten equation**.

When $[S] = K_m$ then $v = V/2$, so that K_m is equal to the *substrate concentration when the rate of reaction is half the maximum rate.* Similarly, this equation predicts that at low substrate concentrations where $[S] \ll K_m$ the rate, v, will be directly proportional to the substrate concentration, while at very high values of $[S]$ where $[S] \gg K_m$ the rate will be *independent* of substrate concentration and *equal to the maximum rate*, V. These features of the variation of the rate of an enzyme-catalysed reaction with substrate concentration are illustrated in Figure 3.7a which is the curve obtained experimentally with many enzymes.

Although the value of K_m can be obtained directly from a plot such as this, it is more usual and accurate to use a linear relationship to determine K_m. Traditionally, the double reciprocal plot, $1/v$ against $1/[S]$, has been used, which gives intercepts on the y and x axes of $1/V$ and $-1/K_m$ respectively. However, for the application of statistical methods of finding the best straight line, other methods such as plotting $[S]/v$ against $[S]$ or v against $v/[S]$ have been used. More recently the preferred method is plotting each pair of values of v and $[S]$ against each other directly when V and K_m can be read directly from the intersection of the straight lines (Fig. 3.7b).

If the concentration of active enzyme is known then, from measured values of V, the value of k_{+2} can be calculated. This rate constant is called the **turnover number** and is defined as the number of substrate molecules transformed in unit time when the substrate concentration is sufficiently high to saturate the enzyme, i.e. the maximum rate of reaction is reached. Values of the turnover number range from 1 to 10^6 per second and are often related to the physiological function of the enzyme. Carbonic anhydrase, responsible for the rapid solution of carbon dioxide in blood, has a turnover number of 6 × 10^5 per second and acetylcholinesterase, which is necessary for the transmission of nervous impulses, is 3 × 10^4. On the other hand, chymotrypsin, involved in the relatively slow process of intestinal digestion of protein, has a turnover number of only 100 per second.

The significance of K_m is rather more complex. At $[S] \gg K_m$ the enzyme will be working at its maximum capacity while at $[S] \ll K_m$ only a few of the enzyme's sites will be filled and the reaction rate will be responsive to the substrate concentration. An example of the operation of these concepts can be seen in the role of hexokinase and glucokinase in liver. Both of these enzymes are able to convert glucose to glucose-6-phosphate, the initial stage in the metabolism of glucose. While hexokinase has a K_m of approximately 10^{-5} M and is able to act on a variety of hexose sugars, glucokinase has a K_m of the order of 10^{-2} M and is specific for glucose. After a carbohydrate-containing meal, the concentration of glucose in the portal vein and hence in the liver may rise to between 5 and 10 mM. As hexokinase is fully saturated by 1 mM, it is unable to respond to this increased glucose by converting it to glucose-6-phosphate. Glucokinase is able to respond since these concentrations are only of the same order as its K_m. Thus while hexokinase ensures the normal smooth introduction of dietary glucose into metabolism, *glucokinase acts as a metabolic safety valve and ensures that circulating blood glucose is kept within fairly narrow limits.*

There is another important difference between hexokinase and glucokinase. Hexokinase is inhibited by glucose-6-phosphate while glucokinase is not. When energy demand is high then the concentration of glucose-6-phosphate is low *but as energy demand falls, the concentration of glucose-6-phosphate increases and inhibits the hexokinase and hence decreases the rate of glucose*

Figure 3.7 (a) The variation of reaction rate with substrate concentration. See text for definition of symbols. The solid and dotted lines represent the behaviour of non-allosteric and allosteric enzymes respectively. (b) Direct plot of reaction rate against individual substrate concentrations to obtain k_m and V from point of intersection.

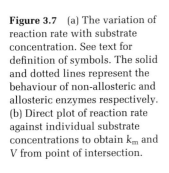
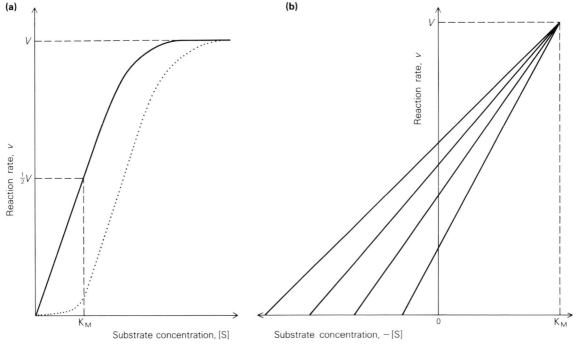

utilisation. In this way the cell is able to respond quickly and accurately to its metabolic needs at any time. The property of many enzymes to respond to changes in concentration of molecules other than substrate is vital to the control of metabolism and, as we shall see in a later section (§3.5), there is a variety of mechanisms by which enzyme activity may be modified.

3.4 Mechanism of enzyme action

3.4.1 CHEMISTRY OF CATALYSIS

Since the discovery by the Buchner brothers that enzymes are capable of being extracted from cells and of functioning in a synthetic, non-cellular environment, it has been implicitly assumed that the role of an enzyme is purely a catalytic one and as such its mechanism of action may be explained on the same basis as catalysis by non-biological catalysts. Similarly, the mechanism of the catalysed reaction must be similar to that of the uncatalysed reaction or at least bear some resemblance to one of the well established organic reaction mechanisms. The tremendous macromolecular complexity of enzymes and the sparsity of firm data concerning their active centres meant that, until comparatively recently, the exact mechanism of action of only a very small number of enzymes had been investigated and no clear pattern of the mechanism of action of enzymes in general had emerged. As more and more enzymes have been studied it has become clear that all enzymes act by *using standard organic reaction mechanisms to effect the transformation of substrate to product*, the role of the enzyme being to provide groups to allow such mechanisms to proceed smoothly and at an increased rate to that which might obtain in their absence. An understanding of these ideas can be gained by considering a few examples.

One of the *most important synthetic reactions in the cell is the formation of a carbon-carbon bond*. By such reactions the small molecules of intermediary metabolism can be used to make a variety of larger molecules, which in turn become the monomeric units for polymer formation. In the test-tube an example of such a reaction is the **aldol condensation** undergone by aldehydes such as ethanal (acetaldehyde).

The reaction is catalysed by a base and may be thought of as taking place in three stages. In the first stage the OH^- of the base removes a H^+ ion from the aldehyde to form the negatively charged **carbanion** which exists as a **resonance hybrid**:

$$OH^- + RCH_2\overset{|}{\underset{H}{C}} = O \rightleftharpoons H_2O + R\overset{|}{\underset{H}{\bar{C}}H} = O \longleftrightarrow RCH = \overset{|}{\underset{H}{C}} - O^-$$

i.e.

$$\overset{\delta-}{RCH} \cdots \overset{}{C} \cdots \overset{\delta-}{O}$$
$$\underset{H}{|}$$

This carbanion with a partial negative charge on the α-carbon atom attacks another aldehyde molecule at the carbonyl carbon atom which has a partial positive charge due to the polar character of the carbonyl bond:

$$
\begin{array}{ccccc}
\delta+ \ \ \delta- & \ \ \delta- & \ \ \delta- & & O^- \\
RCH_2C{=}O \ + \ RCH{-}\!{-}\!{-}C{-}\!{-}\!{-}O & \rightarrow & RCH_2C{-}CHCHO \\
\ \ \ \ \ | & \ \ \ \ | & & \ \ \ | \ \ | \\
\ \ \ \ \ H & \ \ \ \ H & & \ \ \ H \ \ R
\end{array}
$$

The resultant alkoxide ion picks up a proton from water, liberating a OH⁻ ion which is hence reformed and not used up in the reaction:

$$
\begin{array}{ccc}
O^- & & OH \\
| & & | \\
RCH_2CHCHCHO \ + \ H_2O & \rightarrow & RCH_2CHCHCHO \ + \ OH^- \\
| & & | \\
R & & R
\end{array}
$$

Because of its dependence on high concentrations of OH⁻ ions and hence a high pH, the aldol condensation reaction cannot take place as such in the cell.

There is a specific aldolase enzyme which catalyses the linking of two 3-carbon units, pyruvate and glyceraldehyde-3-phosphate, to form a 6-carbon molecule, 2-keto-3-deoxygluconate-6-phosphate:

$$CH_3COCOO^- \ + \ CHO.CHOHCH_2O \ \textcircled{P} \ \rightleftharpoons \ COO^-.COCH_2CHOHCHOHCH_2O \ \textcircled{P}$$

This enzyme acts by a very similar mechanism to that of the aldol condensation (Fig. 3.8). A protonated lysine group serves to bind, forming a **Schiff base**, and to orientate the substrate at the active centre. If the substrate-enzyme mixture is cooled and a suitable reducing agent such as sodium borohydride is added, the C=N is reduced and the reduced complex is stable and can be isolated and characterised. The formation of the complex brings one of the hydrogen atoms of the methyl group near a strong proton acceptor and hence the hydrogen atom is removed in an analogous manner to the action of base in the non-enzymic reaction. This leaves the way open for attack by the carbonyl group of the glyceraldehyde-3-phosphate and the production of the 6-carbon unit, which picks up a proton from a nearby proton donor and finally leaves the enzyme molecule by hydrolysis of the C=N bond.

Chymotrypsin is an enzyme responsible for partial protein digestion in the intestine but is also able to hydrolyse simple esters

$$R_1CONHR_2 \ + \ H_2O \ \rightarrow \ R_1COO^- \ + \ ^+H_3NR_2$$

and

$$R_1COOR_2 \ + \ H_2O \ \rightarrow \ R_1COO^- \ + \ ROH \ + \ H^+$$

Both of these reactions are catalysed by either strong acid or strong alkali in the test-tube. It has been shown that the action of chymotrypsin proceeds in three stages. Firstly, there is binding of the substrate to the active centre of the

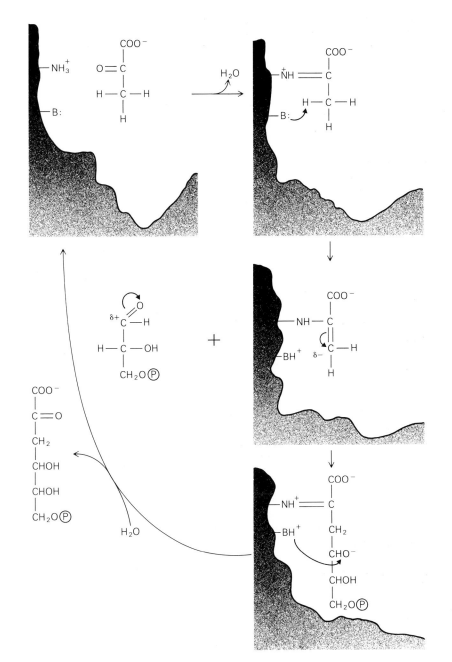

Figure 3.8 Reaction mechanism of the specific aldolase which catalyses the conversion of pyruvate and glyceraldehyde-3-phosphate to 2-keto-3-deoxygluconate-6-phosphate. The shaded portion represents the enzyme-active site with a protonated NH_2 group of lysine and B, a strong proton acceptor.

enzyme, bulky hydrophobic groups fitting into a pocket as shown in Figure 3.5 and hydrogen bonds between CO and NH groups on the enzyme and the substrate stabilising the complex. This brings a specific serine group near the carbon atom of the carbonyl group leading to the second stage, the formation of a transient tetrahedral intermediate (Fig. 3.9). This is only made possible by the transfer of a proton from the serine to a neighbouring histidine, which in turn interacts with a nearby aspartate. The serine-histidine-aspartate inter-

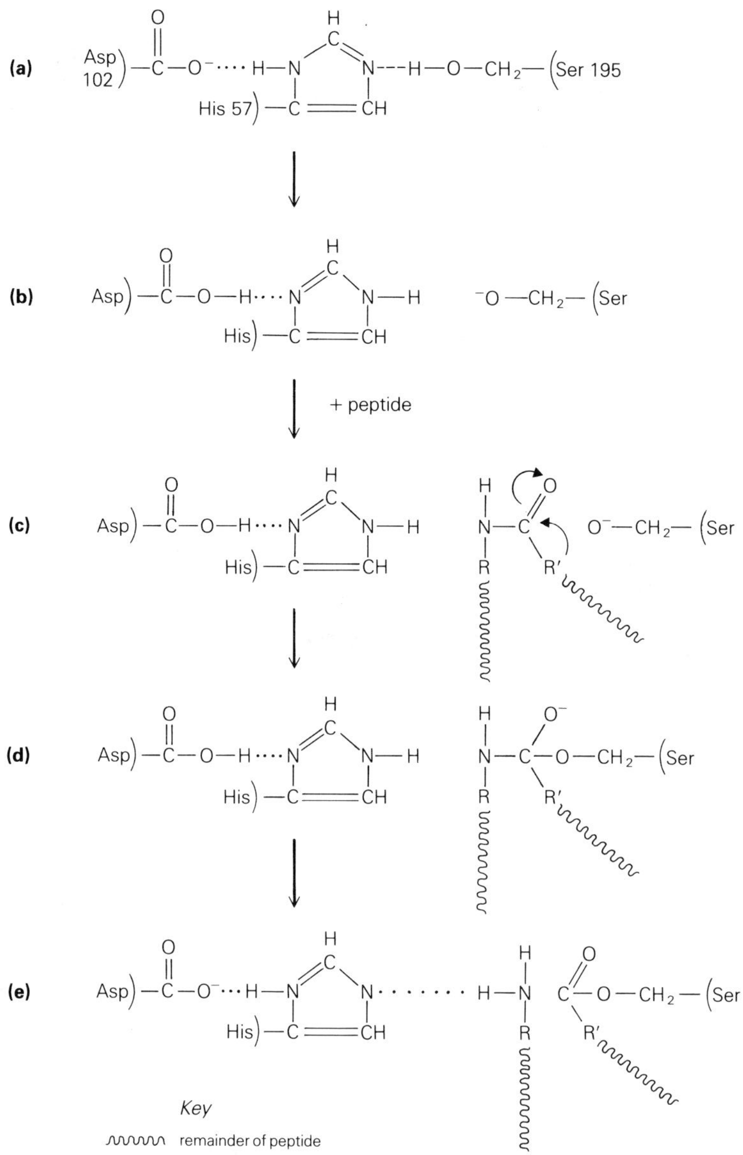

Figure 3.9 The charge relay network in chymotrypsin. Proton transfer in the initial active site (a) leads to deprotonation of a serine group as in (b). Nucleophilic attack by this serine on the carbonyl C atom of a peptide bond (c) gives a tetrahydral intermediate (d). Transfer of a proton from histidine to the N atom of the peptide bond leads to cleavage of the bond (e). For clarity, only the functional groups of the aspartate, histidine and serine in the charge relay network are shown. R and R′ represent the remainder of the peptide molecule.

action is known as a **charge relay network** and serves as a proton shuttle during catalysis. It has the effect of making the OH of serine readily ionisable, a process which does not usually occur. This illustrates a fundamental point in considering the reactivity of various amino acid sidechains in proteins. *Reactivity is a function of the micro-environment of the group, i.e. the nature of the other amino acids that surround it.* In this case we refer to the unique reactivity of the serine group in chymotrypsin. It is certainly different from the other 27 serine residues in chymotrypsin. There is a similar uniquely reactive serine in trypsin, subtilisin and thrombin. The shuttled proton is now added to the nitrogen atom of the peptide bond and allowing cleavage of this

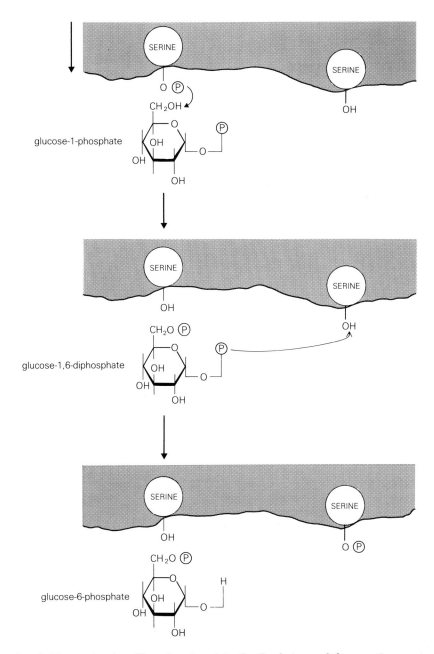

Figure 3.10 Reaction mechanism of phosphoglucomutase showing the alternate phosphorylation of two unique serine groups at the active site (shaded) and glucose-1, 6-diphosphate as an intermediate.

bond. The serine is still acylated and in the final stage of the reaction, water removes the acyl group, giving an acid and restoring the free serine.

Phosphoglucomutase is an unusual enzyme which catalyses an isomerisation reaction: the conversion of glucose-1-phosphate to glucose-6-phosphate. It is hard to imagine any simple organic chemical mechanism for this reaction and yet it is readily accomplished by the enzyme. The secret here is that *the reaction proceeds via the formation of glucose-1,6-diphosphate*. The enzyme contains a phosphorylated serine, again uniquely reactive, and this phosphate is transferred to the 6-position of the glucose molecule. The other phosphate on the 1-position is now transferred to another serine on the enzyme and the whole process can start all over again (Fig. 3.10). If the enzyme-substrate mixture is frozen rapidly it is possible to isolate the enzyme with a phosphorylated serine and hence provide firm evidence of the operation of this mechanism.

3.4.2 ACTIVE SITE IDENTIFICATION

There is now a variety of ways in which the nature of the amino acid residues at the active site can be investigated. One of the earliest techniques involved the use of group-specific reagents. These are reagents which attack specific groups on amino acid sidechains forming covalent bonds and leading to the loss of catalytic activity. For example, thiol groups react with p-chloromercuribenzoate ($ClHgC_6H_4COOH$) with the elimination of HCl and the formation of an S—Hg bond. If the thiol group is a necessary part of the active centre of the enzyme, then such a treatment will lead to the loss of enzyme activity. There are two major problems in the application of this technique. Firstly, the p-chloromercuribenzoate is not completely specific for thiol groups and under some circumstances will react with other sidechains, such as a free amino group in lysine. Secondly, *loss of activity does not necessarily imply modification of the active site*. As knowledge of protein structure has expanded, it has become obvious that minor changes in regions of the enzyme molecule, *distant from the active site*, may lead to sufficiently large conformational changes that the active site is distorted and becomes inactive.

The technique of **differential labelling** is an improved use of group-specific reagents. The enzyme is first reacted with the reagent in the presence of substrate, so that the substrate, being bound at the active site, will protect any reactive groups at the active site from being modified. If the mixture is now dialysed to remove both substrate and excess reagent, it has been shown in many cases that the enzyme retains the bulk of its catalytic activity. The modified enzyme is now allowed to react with fresh reagent, preferably tagged with a radioactive label, and the loss of activity measured as a function of the incorporation of radioactivity. *If the enzyme is now hydrolysed, it is possible to identify which amino acid has been modified by the incorporation of the radio label.* In this way, the lysine group present at the active centre of the aldolase, described previously (see §3.4.1), was first established.

Very recently the lack of specificity of group-specific reagents and their tendency to modify reactive protein sidechains distant from the active centre have been overcome by the introduction of a new technique called **affinity labelling**. An affinity label has two main features. It bears a distinct resemblance to the substrate so that it can be bound specifically at the active site. The other feature is the presence of a 'warhead', a highly reactive group which will react readily with groups in the vicinity of the active site. Chymotrypsin was one of the first enzymes studied by such a technique. It will hydrolyse the substrate tosyl-L-phenylalanine methyl ester

$$C_6H_5.CH_2.CH.COOCH_3$$
$$|$$
$$NHR'$$

where R' is a tosyl group. The aromatic ring serves to direct the substrate to the correct binding site on the chymotrypsin and the enzyme hydrolyses the ester to the corresponding acid and methanol. Tosyl-L-phenyl alanine chloromethyl ketone is an affinity label where the aromatic group is still present to direct binding but there is no ester group to be hydrolysed:

$$C_6H_5.CH_2.CH.COCH_2Cl$$
$$|$$
$$NHR'$$

The highly reactive chlorine is in the same position as the hydrolysed ester bond in the substrate and hence, when the affinity label is bound at the active site, the 'warhead' is free to react with the catalytic group. In the case of chymotrypsin, it was found that the amino acid, modified by the use of the affinity label, was histidine, subsequently shown to be vital to the charge relay network in this enzyme.

In a similar manner, using bromopyruvate $BrCH_2COCOOH$ as an affinity label for the specific aldolase discussed previously, it was shown that the base, which acted as a strong proton acceptor in this enzyme, was the ionised carboyxl group of a glutamic acid residue.

In some cases it is possible to use kinetic data to shed some light on the nature of amino acid residues at an active site. Many groups change their chemical properties between the ionised and un-ionised forms. The carboxyl group can act as a proton donor and can form hydrogen bonds while un-ionised but, on ionisation, it loses both of these properties. Similarly histidine can only accept protons in its unprotonated form. As the state of ionisation of any particular group depends on the pH, it will be apparent that *enzyme activity will vary as a function of pH*. Looking at chymotrypsin again, we can see that the activity variation with pH is a bell-shaped curve (Fig. 3.11). Optimum activity is seen at pH 7.9 and half maximum activities at pH 7 and 8.9. These figures have been interpreted as demonstrating the presence of both histidine and an amino group at the active site. At low pH both of these

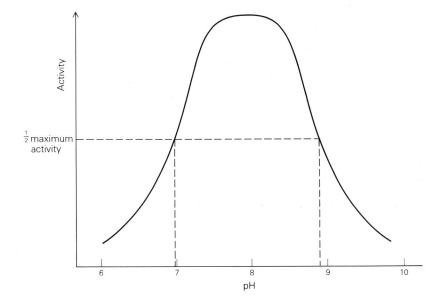

Figure 3.11 The variation of activity of chymotrypsin with pH.

groups are protonated and the enzyme is inactive. As the pH rises, the histidine loses its proton and activity increases. Histidine is known to ionise over the range 5-8 normally in proteins and this fits well with the increase in activity seen between pH 6 and 7.9. Above this latter pH the amino group starts to lose its proton and activity decreases again. Hence the kinetic data show that an unprotonated histidine and a protonated amino group are necessary at the active centre of chymotrypsin for catalytic activity. The histidine we have already seen as part of the charge relay network, and the amino group is part of the binding site that allows correct orientation of the substrate at the active site.

The optimum pH of an enzyme *can often be related to the tissue or cellular distribution of that enzyme.* For instance, pepsin is a proteolytic enzyme which acts in the stomach. Pepsin only has significant activity below pH 2 and thus it is appropriate that the gastric contents should be acidified by secretion of strong hydrochloric acid from the gastric mucosal cells. One might question why this acidification is necessary when there are abundant amounts of other enzymes, such as trypsin and chymotrypsin, available later in the gastro-intestinal tract which are functional at neutral pH. It is known that native, undenatured proteins are poor substrates for these enzymes but that the activity increases markedly if the protein is denatured and susceptible bonds in the protein molecule made more accessible to the enzyme. *One of the simplest ways of denaturing a protein is to subject it to extremes of pH.* The acidity of the stomach causes the protein to unfold and be attacked by the pepsin, which hence must be active at this pH. Breakdown is not extensive in the stomach but the large polypeptides which are formed are *no longer able to form significant secondary structure and are excellent substrates for trypsin and chymotrypsin.*

Similarly, the intracellular digestive enzymes found in the lysosomes have an acidic pH optimum, although in this case the range is between pH 4 and 6. It is possible to show that the pH of a lysosome, actively degrading material, is of the order of 5 so that the property of the enzyme again adequately reflects its environment. In this case it seems likely that the acidic environment and acidic pH optimum is necessary to limit damage to the rest of the cell during the synthesis and packaging of the lysosomal enzymes. When material is provided for the lysosomal enzymes to digest, this is the stimulus for acidification of the particle. Lysosomal proteases are able to act on native proteins and hence denaturation is not a factor in this case.

The majority of enzymes function within the pH range 6-8. Since the variation of activity with pH is determined by the state of ionisation of specific amino acid sidechains at the active site, the property of acting at neutrality would seem to limit relevant amino acids to histidine and possibly tyrosine, both of which ionise in this range. However, it has been found that *in the micro-environment of the protein, ionisation of amino acids takes place at pH values far removed from those normally associated with the particular amino acid.* Carboxyl groups which normally ionise in the range 2-4 may be still protonated at pH 6. Similarly, amino groups may become deprotonated as low as pH 7 instead of in the range 9-11. A partial explanation of this fact is the hydrophobic environment surrounding many active sites. It is common for active sites to be found in a cleft of the enzyme surface, where water molecules may be largely excluded. Since our concepts of ionisation have been developed mainly on the basis of aqueous solution, *hydrophobic environments can produce some very different results.*

3.4.3   FLEXIBILITY: INDUCED FIT AND ALLOSTERIC SITES

Traditionally the interaction of enzyme with substrate is compared with the complementarity of lock and key. There is a great deal of truth in this analogy and we have already seen many examples of specific interaction between three-dimensionally orientated groups of enzyme and substrate. However, it is now realised that many enzymes interact with their substrates by a process known as **induced fit**. In such a process, the shape of the active site is modified by the binding of substrate and only then is the shape complementary to that of the substrate. A good example of this is hexokinase, where the binding of glucose to a cleft in the enzyme surface causes the enzyme to fold around the glucose molecule, forcing any water molecules out. This is important because the phosphate group for the production of glucose-6-phosphate comes from ATP, which can equally well give its phosphate group to water. By drastically reducing the local concentration of water at the active site, this wasteful process is avoided.

Induced fit is possible because of the inherent flexibility of the protein molecule, as discussed earlier in the consideration of protein folding. Flexibility also leads to another phenomenon, **allosterism**. In allosteric enzymes, the binding of a molecule at one location on the surface of the protein can lead to a change in conformation, and hence activity, at another location. Such molecules are called **allosteric effectors** and may be either inhibitors or activators. Allosteric enzymes are often multimeric and the conformational changes are transmitted between sub-units.

Allosteric enzymes show somewhat different kinetics from those observed with other enzymes. The variation of activity with substrate concentration is not the usual rectangular hyperbola, but is instead a sigmoidal curve (see Fig. 3.7a). Two theories have been advanced to explain this behaviour – the **concerted** and the **sequential** model. In their simplest form, each theory assumes that the allosteric enzyme is made up of two identical sub-units, each having one active site. It is postulated that a sub-unit can exist in either a T (tense) or R (relaxed) state, the R state being able to bind both substrate and activator but not inhibitor, while the T state can only bind inhibitor.

In the concerted model the enzyme is postulated to exist in equilibrium between these two forms with the T form greatly in excess. Symmetry is also assumed so that both sub-units in any dimer adopt the same conformation and RT cannot exist. As can been seen in Figure 3.12, the binding of substrate to a R dimer shifts the equilibrium to make more R from T forms. In this way, as the substrate concentration increases, the binding of substrate is made progressively easier, giving the sigmoidal relationship. Inhibitor and activator affect the reaction by stabilising the T and R forms respectively.

In the sequential model, neither symmetry nor equilibrium is assumed. In this model, the initial, rather difficult, binding of a substrate molecule to one sub-unit (T) causes a conformational change in that sub-unit, (to form R) which is transmitted to the other sub-unit, causing a change in its conformation and alteration of its substrate-binding properties. This is quite feasible because we know that sub-units interact specifically at their contact regions and a change in orientation of interacting groups on one enzyme surface can be mirrored by corresponding changes on the other sub-unit surface, which in turn are transmitted throughout the whole of that sub-unit. Activator and inhibitor again act to stabilise the R and T forms respectively.

It is difficult to choose between these two theories. Some enzymes are best explained on the concerted model, others seem to fit the sequential model,

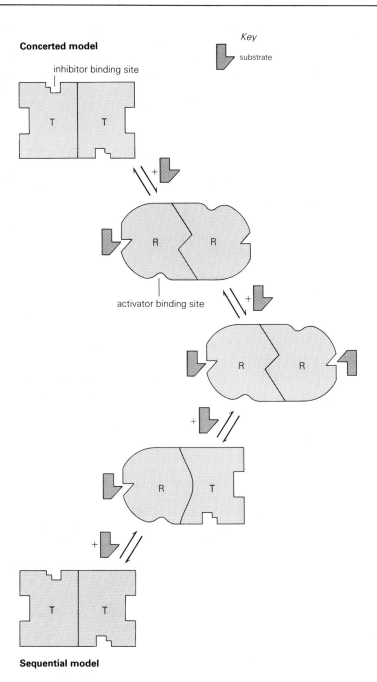

Figure 3.12 The concerted and sequential models of allosteric enzyme interactions. T represents the tense form with inhibitor binding site only and R the relaxed form with substrate and activator binding sites.

while a third group seem far more complex than either of these simple models allows. What is certain is that *allosteric enzymes are vital to the control of cellular metabolism*. By their property of responding specifically and rapidly to molecules other than their substrate, they are able to monitor the metabolic needs of the cell continuously and adjust the level of enzyme activity accordingly.

3.4.4 ISOENZYMES

Many multimeric proteins do not contain a single type of sub-unit but may consist of several different sub-units. In some cases the molecule may consist of a catalytic sub-unit and a regulatory sub-unit, which responds to ligand molecules other than substrate. Catalytic sub-units may combine with a variety of regulatory sub-units, giving a family of enzymes with the same catalytic powers but responsive to different inhibitors and activators. These enzymes often can be separated by electrophoresis (see §1.7) since the charge of the regulatory sub-unit may differ. Such enzymes are frequently referred to as **isoenzymes** and play an important role in the regulation of metabolism.

In other cases an isoenzyme family arises from the aggregation of different sub-units in different proportions. Each enzyme in the family will be a different protein having the *same catalytic function but differing in the ability to perform that function*. Lactate dehydrogenase is an enzyme of universal occurrence in mammalian tissues, but there are subtle differences in the type of sub-unit found in each tissue. In the heart a sub-unit designated H is found, while in muscle and liver a different sub-unit designated M is found. These two sub-units differ in their charge and also in some of their kinetic properties, but the functional significance of these differences to the tissue is still somewhat obscure. The enzyme molecule is normally a tetramer and the full spectrum of possibilities H_4, H_3M, H_2M_2, HM_3, M_4 is found with H_4 predominating in heart and M_4 in muscle. All five forms appear in serum, with H_2M_2 slightly higher than the other four. However, in heart disease some dead cells may release their enzyme contents and this will lead to increased levels of H_4 in serum, providing a useful diagnostic tool for the clinician.

Isoenzymes may also reflect *different stages of maturity* of an enzyme. Enzymes found in the cytoplasm of the cell are synthesised on free ribosomes in the cytoplasm (see §6.8.2) while lysosomal enzymes and proteins for secretion from the cell are formed on ribosomes which are bound to the endoplasmic reticulum. The protein is synthesised vectorially across the membrane of the endoplasmic reticulum, a process dictated by the presence of a signal peptide at the N-terminal end. In the cisternae of the endoplasmic reticulum this peptide is removed and the enzyme protein travels via the Golgi complex to its ultimate cellular or extracellular destination (see §4.2.1). During the course of this transport a complex polysaccharide is added to the protein, forming a glycoprotein. The exact nature of this complex polysaccharide is determined by the primary sequence of the enzyme, and the polysaccharide serves as a label to ensure correct 'packaging' of the enzyme. For example, it seems likely that all lysosomal enzymes contain a phosphorylated mannose residue and that receptors for this sugar in the Golgi complex ensure that all lysosomal enzymes are gathered into one type of particle at this stage in their development. Further modification by the removal of peptide fragments may then proceed, leading ultimately to the mature enzyme. During all these stages the enzyme may show catalytic activity. On electrophoresis, several forms of the enzyme may be demonstrable and these will reflect the differing stages of polysaccharide addition and precursors of the final enzyme.

3.5   Control of enzyme action

Basic metabolism in animals may be thought of as taking place in four stages. In the first stage, nutrient in the diet is broken down into smaller molecules. Dietary protein, polysaccharide and fat is broken down into amino acids, simple sugars and fatty acids respectively. These are then absorbed into the blood stream for transportation to the tissues, where further metabolism takes place. In the second stage, these numerous small molecules are broken down further into a relatively small number of simple units such as acetate, which is normally found in the cell in association with coenzyme A (CoA), forming acetyl coenzyme A. In the third stage, there are two possible fates for these small units. They may either be oxidised completely to CO_2 and H_2O with the liberation of energy, which is stored in the cell as ATP or some other similar high-energy phosphate compound, or they may be used to re-form small monomer molecules such as amino acids, fatty acids or simple sugars. In the final stage, these monomers are converted to macromolecules such as proteins, polysaccharides or fats with specific structural, catalytic or storage roles within the cell. In plants, the nutrient source is simple inorganic material such as CO_2, a nitrogen source and minerals. During photosynthesis these are converted to simple sugars and amino acids, this process replacing the digestion and absorption stage of animals. Subsequent metabolic possibilities are essentially similar to those of animals although the exact structure of storage compounds, e.g. starch and glycogen, may differ.

It is obvious that these various metabolic reactions must be strictly controlled. It would be useless if simple sugars were converted to acetate and oxidised if the energy needs of the cell at that instant were adequately met by existing energy stocks. Far better that the sugar be converted to a storage polysaccharide and kept in readiness for a period of energy need.

The convention of representing metabolism in textbooks as a series of pathways and cycles, while useful in portraying the totality of possible metabolic routes within the cell, is misleading in other ways. Certainly not every enzyme within the cell is acting at its maximum activity at all times. Some enzymes may be functioning at greatly reduced levels, while others may be completely inactive. Many pathways are not uniquely dedicated to the production of one end product from the starting material. Instead there are a variety of branch points on the pathway and the particular metabolic route taken by any molecule must depend on the metabolic needs of the cell at that particular time. Thus sugars can be converted to glycerol, glycogen or starch rather than acetate if the cell needs glycerol for fat synthesis or needs to store reserves. The control mechanisms of the cell are highly complex and, as yet, incompletely understood, but this section will attempt to show some of the basic principles underlying such control.

3.5.1 CONTROL BY SUBSTRATE

This is the simplest form of control. Since the rate of an enzyme-catalysed reaction is proportional at low substrate concentrations to the concentration of substrate, as the concentration of any particular substrate builds up, more of it will be converted to product in an attempt to reduce its concentration. There are many examples of this sort of enzyme. Glucokinase, which we discussed previously, is a typical example. Having a high K_m it usually functions at a rather low rate. As the concentration of glucose builds up, the activity of glucokinase increases to allow the phosphorylation of glucose to proceed at a faster rate. On the other hand there are some enzymes, such as hexokinase, which have a low K_m and are always apparently functioning at their maximum activity. This maximum activity is only apparent, however, since the activity of hexokinase is reduced by the presence of inhibitors.

3.5.2 CONTROL BY INHIBITORS

Inhibitors of enzyme activity may be classified as **reversible** or **irreversible**. The reaction of p-chloromercuribenzoate (see §3.4.2) is an irreversible one. The activity of the enzyme may not be restored by removal of this reagent by dialysis or some similar technique. The reagent has formed a covalent bond with the enzyme and the only possible way that activity could be restored is by chemical cleavage of that covalent bond, a process which itself is likely to denature the enzyme. Irreversible inhibitors *play no part in normal metabolic control* and indeed rarely occur naturally. They have played a significant part in the design of pesticides and herbicides for, by completely and irreversibly inhibiting some specific enzyme in an organism, it is possible to wipe out that enzyme species over a confined area.

Reversible inhibitors may be divided into three main types: **competitive**, **non-competitive** and **allosteric**. Competitive inhibitors act by combining with the active site of the enzyme preventing access by substrate. Because of their competitive nature, their structure is often very similar to that of the substrate and the activity of the enzyme will depend on the relative concentrations of substrate and inhibitor and the relative binding affinities of these molecules for its active site. The enzyme diphosphoglycerate mutase is an important enzyme in the metabolism of glucose in red blood cells and catalyses the interconversion of 1,3-diphosphoglycerate and 2,3-diphosphoglycerate:

$$\begin{array}{ccc}
\text{COO}\ \textcircled{P} & & \text{COO}^- \\
| & & | \\
\text{HO—C—H} & \rightleftharpoons & \textcircled{P}\,\text{O—C—H} \\
| & & | \\
\text{CH}_2\text{O}\ \textcircled{P} & & \text{CH}_2\text{O}\ \textcircled{P}
\end{array}$$

It is competitively inhibited by low concentrations of 2,3-diphosphoglycerate and hence if this metabolite builds up, the interconversion is slowed. It is common *for the product of a reaction to be a competitive inhibitor of the substrate because of their structural similarity.* Metabolically, this makes a great deal of sense, since the implication of increasing product level is that it is being formed faster than it can be used up in subsequent reactions of the pathway and it is wasteful to form more in an unregulated manner.

Non-competitive inhibitors are equally important in metabolism. This type of inhibitor can bind to the enzyme simultaneously with the substrate so that there is no element of competition. It seems likely that *most non-competitive inhibitors modify the catalytic site rather than the binding site*. Hexokinase shows non-competitive inhibition with glucose-6-phosphate. If the concentration of glucose-6-phosphate, the production of the reaction, builds up, the activity is inhibited; this inhibition is independent of the prevailing concentration of glucose. Only if further metabolism of glucose-6-phosphate takes place and the level of this metabolite decreases, can phosphorylation of glucose proceed again normally.

3.5.3 CONTROL BY ALLOSTERIC EFFECTORS AND ISOENZYMES

If one considers a reaction pathway such as

$$A \rightarrow B \rightarrow C \rightarrow D \ldots X \rightarrow Y \rightarrow Z$$

there are at least two ways in which inhibitors can exert metabolic control. As the end product builds up, it can inhibit its production from Y, which in turn will build up and inhibit its production from X. Ultimately, the concentration of D will rise and inhibit $C \rightarrow D$ until finally the conversion of A to B is inhibited and the whole pathway comes to a standstill, only to proceed again when the concentration of Z falls. Although such a system works, it is rather wasteful and slow. Each of the intermediates on the pathway *must build up to inhibitory levels and this is both costly in raw materials and time-consuming*. This problem is overcome by certain pathways and enzymes which show **feedback inhibition**. In this type of inhibition one of the early reactions in the pathway, often the first, is inhibited by the end product. In our example, the conversion of A to B would be inhibited by high concentrations of Z. Feedback inhibition provides a rapid and efficient mechanism for switching off a pathway when the material that it operates to produce, accumulates, without wasteful accumulation of all the intermediate compounds. One of the end products of the metabolism of glucose can be thought of as ATP. When the energy needs of the cell are being adequately met and hence any ATP produced is being stored, it would be advantageous to divert glucose to some other use. One way of doing this would be by inhibiting its phosphorylation to form glucose-6-phosphate. Hexokinase is inhibited by ATP and in this case the ATP functions as an allosteric inhibitor. The structure of ATP is dissimilar to that of glucose and it binds to a regulatory site on the hexokinase.

A rather more complex example is seen in the synthesis of certain amino acids by bacteria. Aspartate can be converted to lysine, methionine, threonine and isoleucine by a complex branched pathway (Fig. 3.13). It is important that each of these amino acids is present at high enough concentrations for the bacterium to make the various proteins that it requires and yet, since nitrogen is often in short supply for bacteria, excessive production of any amino acid would be deleterious. Consider firstly the production of isoleucine from threonine. Here we have a simple feedback inhibition such that the conversion of threonine to α-ketobutyrate is allosterically inhibited by isoleucine. Similarly, lysine allosterically inhibits its first reaction from aspartate semialdehyde. However, the build-up of lysine also means that there is less production required of aspartyl phosphate from aspartate by aspartokinase. If this enzyme was completely inhibited, production of methionine and threonine and isoleucine would suffer. It has been found that there are at least three isoenzymes of aspartokinase and one of these is allosterically inhibited by lysine. On a similar basis another of the isoenzymes is inhibited by threonine. This isoenzyme pattern of aspartokinase is represented on Figure 3.13 by three arrows, two of which show negative feedback inhibition. The use of multiple forms of an enzyme, i.e. isoenzymes, *can successfully solve the problem of control of complex pathways.*

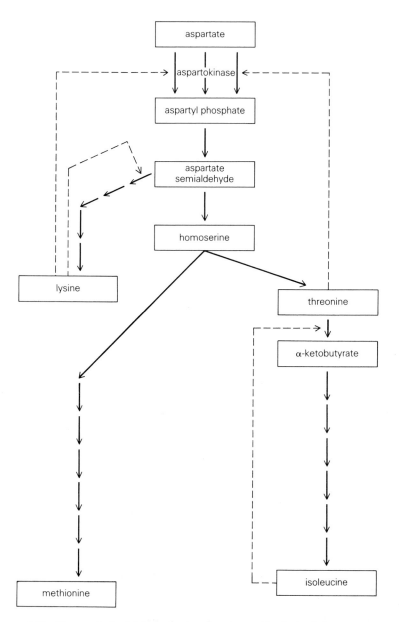

Figure 3.13 The control of biosynthesis of amino acids derived from aspartate in bacteria. Only important intermediates at control points are shown, with other intermediates being represented by arrows on solid lines. Dotted lines show some of the feedback inhibitions of these pathways.

3.5.4   CONTROL BY COVALENT MODIFICATION: CASCADES

Some enzymes exist in the cell as inactive precursors. To release the latent activity of these enzymes, it is necessary to remove a small peptide from the precursor form, called a **zymogen**. Both the digestive enzymes, trypsin and chymotrypsin, exist in the cell as zymogens, trypsinogen and chymotrypsinogen. Storing the enzymes in the cell in this inactive form prevents the possibility of *cellular damage occurring by hydrolysis of intracellular proteins*. Trypsinogen is converted to trypsin, and hence activated, by an enzyme called enterokinase, produced by the cells that line the duodenum. When trypsinogen leaves the pancreatic cell and is secreted into the duodenum, it is activated by proteolytic cleavage by the enterokinase. Once a small quantity of active trypsin is produced, this in turn activates more trypsinogen and the whole process is **autocatalytic**. Chymotrypsinogen is converted to chymotrypsin by trypsin and in this way the active enzymes, necessary for protein digestion, are produced.

There are a variety of other enzymes which are normally present as zymogens and for which activation is a necessary and carefully controlled process. An excellent example of this is seen in blood clotting. The final stage in blood clotting is the conversion of the zymogen, fibrinogen, to fibrin. While fibrinogen exists as discrete protein molecules, fibrin molecules aggregate, forming a three-dimensional network of fibres which traps blood cells and hence plugs any damaged capillaries. Fibrinogen is converted to fibrin by the enzyme thrombin, which in turn exists in a precursor form, prothrombin. Some of the complexity of the system is illustrated in Figure 3.14. At each

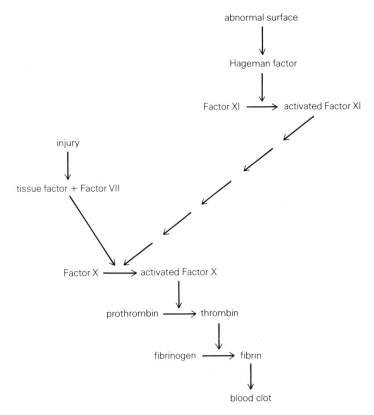

Figure 3.14 The 'cascade' reactions leading to blood clotting. Both the intrinsic (resulting from contact with an abnormal surface) and extrinsic (resulting from tissue damage) pathways are shown.

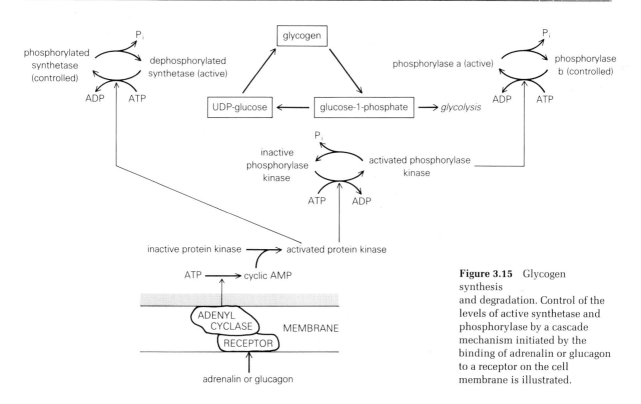

Figure 3.15 Glycogen synthesis and degradation. Control of the levels of active synthetase and phosphorylase by a cascade mechanism initiated by the binding of adrenalin or glucagon to a receptor on the cell membrane is illustrated.

stage an enzyme acts to convert a zymogen to its active enzyme. The net effect is termed a **cascade**, for a few molecules of tissue factor, released in response to injury, or a few molecules of Hageman factor, released by contact with an abnormal surface, can be amplified into the production of many millions of molecules of thrombin and hence fibrin. At each stage in the process *a single enzyme molecule can effect the production of many hundreds of molecules of active enzyme at the next stage.*

Not all covalent modification of enzymes proceeds via the proteolytic cleavage of a zymogen. In some cases, the difference between active and inactive enzyme is the presence of a phosphate on a serine in the enzyme. The pathways for the synthesis and degradation of glycogen are summarised in Figure 3.15. To achieve maximum control there are separate enzymes involved in synthesis and degradation, glycogen synthetase and phosphorylase, respectively. Both of these enzymes exist in two forms: with and without phosphate. The two forms of each enzyme show very different control characteristics. The phosphorylated form of glycogen phosphorylase (called phosphorylase a) is always active and is unresponsive to the levels of other metabolites in the cell. On the other hand, the non-phosphorylated form (phosphorylase b) is active only in the presence of high concentrations of AMP and is inhibited by high levels of ATP. AMP is formed by the breakdown of ATP and is indicative of low levels of energy storage. If the energy stores of the cell are depleted, then energy reserves in the form of glucose produced from glycogen must be mobilised as quickly as possible. On the other hand, when energy stocks are high and ATP concentrations adequate, it is important that glycogen phosphorylase is inhibited and that the stores of glycogen are not broken down. In resting muscle nearly all enzyme is in the controlled form, phosphorylase b.

The reverse argument is true for glycogen synthetase. The dephosphorylated form is always active while the phosphorylated form is only active in the presence of high concentrations of glucose-6-phosphate. The enzyme is present in resting muscle mostly in the dephosphorylated form and is thus able to convert any free glucose-1-phosphate (and hence free glucose) in the cell to glycogen for storage. The interconversion of *the phosphorylated and dephosphorylated forms of both enzymes are catalysed by the same phosphatase in one direction and the same kinase in the other.* Thus, if breakdown is necessary, phosphorylation of both enzymes takes place and this has the effect of stimulating the phosphorylase to uncontrolled activity and reducing the synthetase to a state where it is only active if excessive amounts of glucose-6-phosphate build up.

3.5.5 CONTROL BY HORMONES

All of the mechanisms for control that have been described so far in this section have been concerned with control at the cellular level. The mammalian body is not just a collection of cells or even of tissues. It is capable of acting in concert to a variety of internal and external stimuli. To allow this to happen, the control of individual cells and tissues must be co-ordinated. This is achieved by the use of chemical messengers or hormones which circulate between the tissues. For example, under conditions of stress or low circulating blood sugar concentrations, the glycogen store is mobilised through the influence of the hormones adrenalin and glucagon. Adrenalin acts in stress, and by decreasing the stored glycogen and thus increasing the free sugar levels, it makes more energy available for the body to withstand the stress. On the other hand, if the level of circulating glucose decreases under normal circumstances, glucagon is secreted and this acts again by increasing glycogen breakdown and restoring the circulating glucose concentrations to normal levels. This is important since a decrease in the level of glucose being transported to the brain leads to irreversible brain damage.

Although they influence the rate of glycogen breakdown, neither adrenalin nor glucagon is able to enter the cell. They act by combining with a specific protein in the plasma membrane of the cell, called a **receptor**. This in turn is stimulated to interact with a protein on the inner surface of the plasma membrane. This latter protein, adenyl cyclase is activated and catalyses the production of cyclic AMP (cAMP) from ATP (Fig. 3.15). This in turn starts a cascade of reactions (cf. blood clotting), ultimately activating phosphorylase kinase, which phosphorylates the synthetase and phosphorylase leading to cessation of glycogen synthesis and initiation of glycogen breakdown. In this way, by specific interaction with receptor proteins in the plasma membrane, *some hormones are able to control the rate of intracellular reactions.*

3.6 Trends in research

3.6.1 IMMOBILISED ENZYMES

It is often written that enzymes, being catalysts, are unchanged in the reaction. In practice, very rarely can enzymes be recovered in an active form from a reaction mixture. While in the research laboratory this may be relatively unimportant, in industry, and even in a clinical analytical laboratory, as enzymes find increasing use as reagents, long life and effective recovery become vital. The potential for the application of enzymes to industrial processes is enormous. Their specificity and the fact that they can catalyse reactions to proceed at excellent rates and under mild conditions could make their use an economic necessity.

One of the most interesting breakthroughs in recent years has been the development of immobilised enzymes. By fixing enzymes to an inert, and usually insoluble, support or by encapsulating them in some polymeric gel, the stability and recoverability of the enzymes has been greatly enhanced and the possibility of industrial use has become commercially feasible. There is a variety of ways in which immobilisation may be achieved. The enzyme may be covalently linked to some inert support, often cellulose or some synthetic polymer such as nylon. Provided that the enzyme is linked at some site distant from the active centre, full enzyme activity is often retained. Alternatively, the enzyme may be trapped in a gelatin capsule by forming the gel in the presence of the enzyme. Such a preparation is often very stable but this type of immobilisation is only suitable in cases where both the substrate and product of the reaction are small enough to diffuse freely in and out of the gel particle. An ingenious approach has been to impregnate filter paper or pulp with the enzyme and then to cross-link molecules of enzyme together, effectively entrapping them within the fibres of cellulose. Yet another approach has involved the electrostatic interaction between charged enzyme molecules and an oppositely charged support, such as an ion-exchange cellulose. This technique has the advantages of simplicity and cheapness but great care has to be taken that the conditions under which such a preparation is used do not cause any weakening of the electrostatic bond and hence loss of enzyme. The choice of method at the moment is a purely pragmatic one and depends entirely on the nature of the application envisaged.

Glucose in biological fluids is commonly measured in clinical biochemistry laboratories. For many years, the method used depended on the reducing properties of glucose and assumed that any other reducing sugars were present in insignificant amounts. The use of glucose oxidase, an enzyme that specifically oxidises glucose and is without effect on other reducing sugars, such as galactose, greatly improved the validity of the determinations. However, glucose oxidase is relatively expensive and rather unstable, making it necessary to monitor the activity of the enzyme at frequent intervals. Much analysis in clinical biochemistry laboratories is now fully automated. Measured samples for analysis are pumped with reagent through heated coils and eventually through spectrophotometer cells where the absorbance of each sample is recorded, often on a computer. It has been possible to immobilise the enzyme on the nylon tube through which the sample passes and, by adjusting the flow rate, to ensure complete reaction during the passage of the substrate through the tube. The possibilities for the development of this type of approach are endless.

There are many people, particularly among those whose ethnic origin is African, who lack the enzyme lactase. Since most animal milks contain lactose, they are denied a rich source of nutrient. If fed milk, lactose is excreted because it is not broken down to glucose and galactose in the intestine and this, in turn, causes diarrhoea. Yeast contains a powerful β-galactosidase, the enzyme that hydrolyses lactose. It is possible to extract this enzyme from yeast and prepare it in a highly purified state. The enzyme is entrapped in filter beds and the milk allowed to flow through the beds. A major plant in Italy is now capable of handling 30 000 litres of milk a day using 500 g of entrapped enzyme, the enzyme needing to be renewed only twice a year. In this way large volumes of glucose can be produced relatively inexpensively.

Amino acids to supplement diets or for the production of purely synthetic diets are now used in large quantities. It is relatively easy to synthesise amino acids chemically but the product is a **racemate**, i.e. it contains equal amounts of the D and L stereo-isomers. As the body can only use the L form, half the product is wasted and often has to be removed. The amino acid can easily be converted to its acyl derivative. There is an enzyme, aminoacylase, which can be prepared from moulds and which hydrolyses only L-acylamino acids. The unhydrolysed D-acylamino acid is then treated with a racemase which converts 50 per cent of it to the L form, and the whole cycle is repeated. After several turns of the cycle, virtually all of the product has been recovered as the L-amino acid. On a commercial scale and using columns containing enzyme electrostatically bound to ion-exchange celluloses, it is possible to produce up to 800 kg of amino acid per hour. *Industrially, it seems that the application of immobilised enzymes is limitless and it is only cost effectiveness that at present governs their usage.*

3.6.2   INBORN ERRORS OF METABOLISM

There is a large variety of human and animal diseases which are genetically determined and hence can be transmitted from generation to generation. If both the parents of a particular child have one mutant allele of a specific gene, then there is a one in four chance that the child will inherit genes where both of the alleles are mutant. This manifests itself in the production of a mutant protein, often an enzyme. In some cases the mutant gene has a nonsense code and there is complete failure to synthesise the specified protein, while in other cases an enzyme with greatly reduced activity or an enzyme which is metabolically unstable and is rapidly degraded in the cell is produced. If the enzyme is vital to the metabolism of the cell, the disease will be fatal, and the child will die in infancy. At the extreme some enzyme deficiencies are so lethal to the cell that the foetus fails to develop normally and spontaneous abortion may result. Certain enzyme deficiencies, while resulting in recognisable physical defects, have very little influence on the development of the individual and a nearly normal life is possible. Thus people suffering from albinism have a defect of the enzyme tyrosinase, and this accounts for their lack of pigmentation, a condition which may cause a certain amount of embarrassment and even minor side effects such as intolerance of strong sunlight, but allows the individual to lead a normal and healthy life.

In many cases it is possible to identify the specific enzyme which is deficient in a particular disease. For example, there is a group of diseases, collectively termed the *lysosomal storage diseases*, where a specific lysosomal enzyme is missing. The lysosomes are organelles where the intracellular digestion of the cell takes place. The material for digestion may be of extracellular origin, such as virus particles or bacteria, or may be parts of the cell, such as mitochondria or other membranous components which have become 'aged' and which it is necessary to break down in order to return their amino acids, sugars, nucleotides and fatty acids for use in the synthesis of new macromolecules. The complexity of macromolecules necessitates the use of a battery of enzymes for complete degradation, and if any of these enzymes is missing then degradation is halted at that point. The undegraded material accumulates in the lysosome and this interferes with the normal functioning of other lysosomal enzymes until gross cellular malfunction is observed. The patient suffers severe skeletal changes and often a failure of normal mental and locomotor development, and survival beyond infancy is rare. Sometimes the enzyme is not missing but reduced to about 10 per cent of normal. In such cases the deficiency may not manifest itself until late childhood or post-puberty. Perhaps this is not surprising as the parents are heterozygous and survive perfectly normally on 50 per cent of the normal enzyme activity.

A good example is a group of lysosomal storage diseases called the **sphingolipidoses**. Sphingolipids are complex glycolipids which are an integral and important part of all cell membranes (see §4.1.1). They are particularly concentrated in membranes of nervous tissue and in the brain. The sequential removal of sugars from a sphingolipid molecule is shown in Figure 3.16, together with the enzyme catalysing the removal until the basic lipid, ceramide, is reached. In all cases, a specific disease associated with the

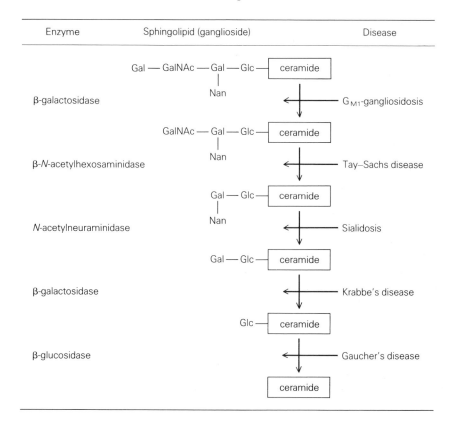

Figure 3.16 Sphingolipid degradation in the lysosome. The sequential action of individual enzymes and the diseases arising from specific enzyme deficiencies are shown. Glc, Gal, GalNAc and Nan represent glucose, galactose, N-acetylgalactosamine and N-acetyl neuraminic acid respectively.

deficiency of a single enzyme has been identified and, although the incidence of these diseases is not great (between 1 in 25 000 and 1 in 200 000 live births for a particular disease), the cumulative effect of over 100 such diseases and the individual anguish caused are matters for concern.

The role of clinical biochemists in the management of such diseases is twofold. They must provide rapid and completely reliable evidence of the enzyme deficiency and they must attempt to develop techniques of enzyme therapy. The first function is now adequately fulfilled in most cases. Highly sensitive assays of enzymes, using fluorescent or radioactive labels (see §1.3.3 and §1.8.4) have made deficiency diagnosis possible using a small number of cells and even, in some cases, a single cell. Since the amniotic fluid bathing the foetus in the uterus contains cells shed from the skin of the foetus, by the collection of amniotic fluid, a process called **amniocentesis**, and the growth of the cells in culture tubes, it is possible to detect lysosomal storage diseases in the developing foetus and to provide reliable information on which genetic counselling may be given.

Enzyme replacement therapy has been far less successful. It is possible to prepare pure enzyme from a variety of sources, often microbial. The difficulties lie in presenting the enzyme to the patient in an acceptable form and in ensuring that the enzyme is targeted to those tissues where it is most needed. Rejection on antigenic grounds has, to some extent, been overcome by enclosing the enzyme in a lipid bilayer called **liposome**. Targeting is far more difficult and as yet it has been impossible to cross the blood-brain barrier and allow the enzyme to enter the brain, the tissue often most affected. In a few cases where the disease has been mostly skeletal in nature it has been possible to achieve some amelioration of the disease by therapy but the general application of this technique is a long way off. Indeed it is possible that recent advances in genetic engineering may provide a more suitable therapy. What is certain is that our knowledge of the mechanism and control of enzymically catalysed reactions has allowed many diseases to be explained on a molecular basis. The use of gene manipulation in the prenatal diagnosis of disease is discussed in Section 6.10.3.

References and further reading

BOOKS

Alberts, B., D. Bray, J. Lewis, M. Raff, K. Roberts and J. D. Watson 1983. *Molecular biology of the cell*. New York: Garland.

Armstrong, F. B. 1983. *Biochemistry*, 2nd edn. Oxford: Oxford University Press.

Stryer, L. 1981. *Biochemistry*, 2nd edn. New York: W. H. Freeman.

SELECTED ARTICLES

Changeux, J.-P. 1965. The control of biochemical reactions. *Scient. Am.* **212** (Apr.), 36.

Dayhoff, M. O. 1969. Computer analysis of protein evolution. *Scient. Am.* **221** (July), 86.

Dickerson, R. E. 1972. The structure and history of an ancient protein. *Scient. Am.* **226** (Apr.), 58.

Dickerson, R. E. 1980. Cytochrome c and the evolution of energy metabolism. *Scient. Am.* **242** (Mar.), 98.

Friedmann, T. 1971. Prenatal diagnosis of genetic disease. *Scient. Am.* **225** (Nov.), 34.

Koshland, D. E. 1973. Protein shape and biological control. *Scient. Am.* **229** (Oct.), 52.

Kretchmer, N. 1972. Lactose and lactase. *Scient. Am.* **226** (Oct.), 52.

Mosbach, K. 1971. Enzymes bound to artificial matrixes. *Scient. Am.* **224** (Mar.), 26.

Stroud, R. M. 1974. A family of protein-cutting proteins. *Scient. Am.* **231** (July), 74.

Cell Biology

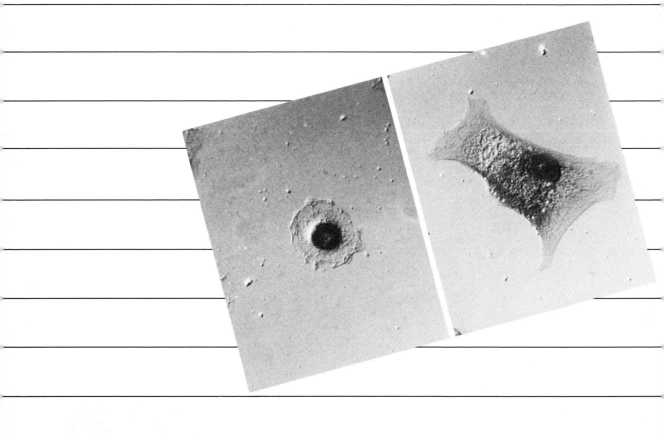

Membranes

G. E. Jones

MODERN VIEWS IN BIOLOGY

4.1   Membrane structure

In recent years, interest in membranes has centred on molecular structure and functional organisation. Research in this subject, involving the combined efforts of virtually every scientific discipline, has revealed a considerably more complicated picture of membranes than could have been imagined in the 1960s when the electron microscope emphasised the abundance of membranes in cells.

Membranes consist mainly of lipid, protein and carbohydrate, together with water which usually makes up about 20 per cent of the total weight. The proportions of the organic components vary from the extremes of myelin, with 80 per cent lipid, to the inner membrane of mitochondria, with 25 per cent lipid. Most plant and animal membranes approach a distribution in which lipid makes up from 30 to 50 per cent of the total membrane dry weight with proteins forming the bulk of the remainder. Carbohydrate content is usually less than 5 per cent of the dry weight, though carbohydrates are more abundant in the plasma membranes of eukaryotic cells.

4.1.1   MEMBRANE LIPIDS

The lipid fraction of biological membranes is highly diverse, both in the total amount of lipid which is present and in the composition of the various lipid classes. *Even different organelles in the same cell can show considerable variation.* The functional significance of the differences in lipid composition is not clearly understood, neither is the molecular basis for the generation of lipid differences between the various membrane systems. All the major lipids share common features which suit them to a bilayer organisation. They are all **amphipathic**, i.e. they have **hydrophobic** and **hydrophilic** domains. The polar (hydrophilic) groups which consist of charged or **zwitterionic** headgroups may face the aqueous exterior while the hydrophobic regions associate with each other in the interior of the bilayer. The most common of the membrane lipids are the phosphoglycerides (Fig. 4.1). All are built upon a glycerol skeleton where one of the three carbons of glycerol (3-carbon) is linked to a phosphate group. This phosphate group is in turn linked to one of a variety of alcohols of which choline and ethanolamine are the most frequently occurring types (Fig. 4.2a). The remaining two glycerol carbons are bound by ester linkages to long hydrocarbon chains derived from fatty acids.

The sphingolipids are similar in structure to the phosphoglycerides but differ in that glycerol is replaced by sphingosine as the core molecule (Fig. 4.2b). Sphingosine is a nitrogen-containing alcohol with an amino (—NH) and

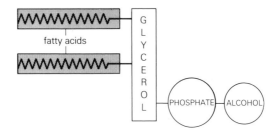

Figure 4.1   Structure of a phosphoglyceride Ø (shaded regions are hydrophobic).

(a)

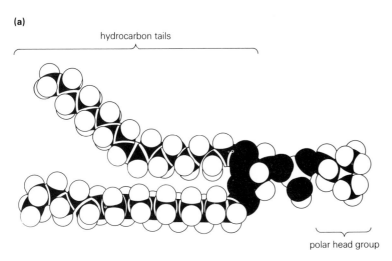

hydrocarbon tails

polar head group

(b)

Figure 4.2 (a) Model of a phosphatidyl choline molecule constructed from space-filling atoms. (b) Sphingosine molecule. Linkage to other groups occurs at the hydroxyl and amino groups.

$$CH_2OH$$
$$H_3\overset{+}{N}-C-H$$
$$OH-C-H$$
$$CH$$
$$\|$$
$$CH$$
$$(CH_2)_{12}$$
$$CH_3$$

Figure 4.3 A glycolipid. Such molecules are a major component of cell membranes.

a hydroxyl (—OH) site available for linkage to other chemicals. The amino group is linked to a long hydrocarbon chain derived from a fatty acid while the hydroxyl group is usually linked to a carbohydrate group or (via a phosphate group) to one of a variety of molecules such as the alcohols choline or ethanolamine. One class of sphingolipids (the ceramides, §3.6.2) have unbound hydroxyl groups.

The major membrane glycolipids of plant and animal cells are made up from sphingolipids with carbohydrate groups attached (Fig. 4.3). They are restricted in distribution, being confined largely to the plasma membranes where they are orientated with the carbohydrate groups extending out into the aqueous phase of the intercellular spaces (Fig. 4.4). The carbohydrate groups are usually 6-carbon sugars such as glucose, mannose, fucose or galactose. Despite the limited range of sugar types, an almost endless variety of glycolipid structures is conferred upon the membrane *by virtue of the different combinations of sugar residues which are possible, as up to 15 sugar units may assemble in branched chains at the polar end of the sphingolipid.*

All sterols are based on a skeleton of four carbon rings (Fig. 4.5). They are almost completely non-polar and are therefore only slightly amphipathic. The major sterol of animal membranes is cholesterol, which may make up to 30 per cent of the lipid content of plasma membranes. Plant membranes *have little or no cholesterol but large quantities of related sterols.* Sterols of any kind are rare or absent in bacterial and blue-green algal membranes (see §2.3).

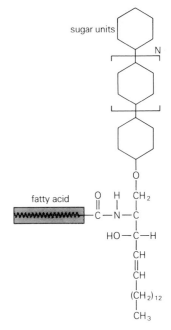

sugar units

fatty acid

Figure 4.4 Sugar residues of glycolipids and glycoproteins are found on the outside surface of the plasma membrane.

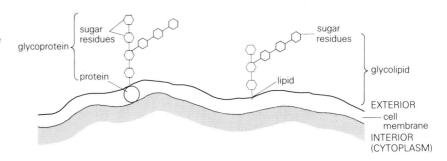

The first development of the concept that membrane lipids were organised as bilayer structures came in 1925 from the experiments of Gorter and Grendel. They found that the lipid extracted from erythrocyte membranes was twice the amount that they had predicted from calculations of the surface area of the cells. Since erythrocytes contain little membrane that is not surface, they concluded that the lipid molecules of the membrane must be organised as a double layer (Fig. 4.6). Even though several faults were later found in the execution of these experiments, they laid the foundation for future models of membrane structure.

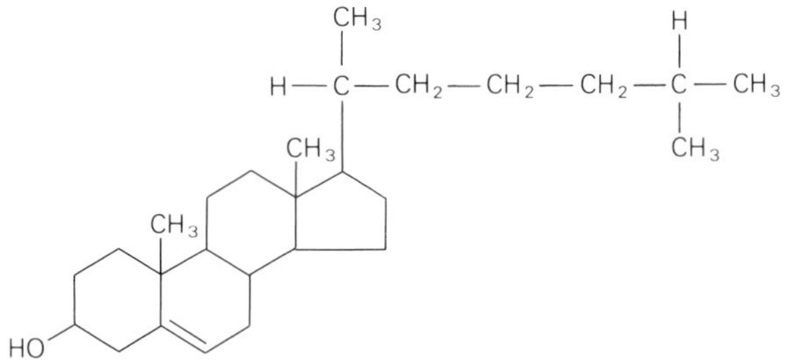

Figure 4.5 Cholesterol – sterol component of animal membranes.

We now know that the hydrophobic nature of lipids makes them tend to aggregate in an aqueous environment to form bilayers in a self-assembly process. More recently, it has been recognised that *individual lipid molecules can move (diffuse) within the membrane*. Using artificial bilayers containing chemically tagged lipids, it has been shown that two types of lipid movement exist. Lipid molecules readily exchange places with their neighbours on one side of the bilayer (more than a million times a second) but they only occasionally migrate from one side of the bilayer to the other. This process, often called a **flip-flop**, happens about once a fortnight for any individual lipid. Lipid behaviour in true biological membranes is generally the same as in artificial bilayers; the lipid fraction of a membrane can thus be thought of as a rather *viscous liquid in which molecules move rapidly by lateral diffusion but are largely restricted to their own side of the bilayer.*

The lack of exchange between the two halves of a bilayer (termed **monolayers**) serves to sustain a second feature of membranes. The lipid compositions of the two monolayers are different: *glycolipids are found only in the external-face monolayer and choline-containing lipids are also more abundant on this face. On the other hand, lipids with terminal amino groups* *are more common in the inner monolayer.* This lipid asymmetry must have a function, though at present we only have a poor understanding of what this might be. However, we do have some idea as to why cholesterol is so abundant in plasma membranes while being relatively scarce in the intracellular membranes of cells. Cholesterol molecules are orientated in the bilayer with their hydroxyl groups close to the polar head groups of other lipids. This provides some mechanical stability and rigidity to the bilayer and *plays a vital role in reducing the fluidity of the plasma membrane at physiological temperatures.* Mutant cells which cannot synthesise cholesterol lyse and die unless provided with the sterol.

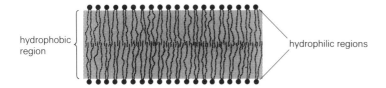

Figure 4.6   Bilayer model of membranes.

4.1.2   MEMBRANE PROTEINS

While the bilayer organisation of lipid confers ordered structure to biological membranes, it is the protein that is responsible for functional specificity. Each type of cellular membrane has its own characteristic complement of proteins, reflecting the metabolic roles played by the various membrane systems of cells. Membranes involved in complex energy transduction such as the inner mitochondrial membrane consist of approximately 75 per cent protein, whereas myelin (used to insulate nerve fibres) contains less than 25 per cent protein. Membrane proteins fall into two major classifications based upon the ease with which they can be isolated from the bilayer. Some can be released by gentle procedures such as changing the ionic strength of the bathing solution; others require virtual destruction of the bilayer with detergents. The former are termed **peripheral** or **extrinsic proteins** and the latter **integral** or **intrinsic proteins**. Because of the difficulties encountered by scientists when trying to isolate the integral proteins, it was only recently that a picture of the organisation of membrane proteins became clear. Our ideas are largely based upon data gained on the plasma membrane of mammalian erythrocytes because this is the most abundant source of good membrane preparations available.

Separation of erythrocyte membrane proteins using gel electrophoresis (see §1.7) has led to the discovery of three major proteins which have since been studied in detail. Approximately 12 other proteins have also been isolated but very little is known about the structure or role of this group.

It so happens that the three major proteins, called spectrin, band III and glycophorin, are arranged in the membrane in different ways; they thus provide us with a good example of the organisation of proteins which *serves for all membranes so far studied.*

Spectrin, a high molecular weight, fibrous molecule, can be removed from the membrane by reducing the ionic strength of the bathing solution. It can therefore be regarded as an extrinsic or peripheral membrane protein. Spectrin is composed of two polypeptide chains (approximately 240 000 and

220 000 **daltons** each) arranged as a network or mesh on the cytoplasmic surface of the membrane. It is thought that spectrin is *largely responsible for maintaining the characteristic biconcave shape of erythrocytes.*

Band III is a transmembrane protein, that is to say it crosses the lipid bilayer of the membrane. Much of the mass of this protein is associated with the lipid phase and detergent treatment is necessary to extract it. Band III is therefore defined as an intrinsic or integral membrane protein. Its polypeptide chain *appears to cross the membrane several times in a folded conformation.* The protein seems to be a dimer and has a small amount of carbohydrate attached to the molecule; as usual the carbohydrate is only found on the external face of the surface. The role of band III seems to be that of an *anion channel in the membrane.* As erythrocytes pass through the lungs, they exchange HCO^- for Cl^- during the process of CO_2 release. Band III acts as a hydrophilic channel which allows these two anions to pass across the membrane.

Glycophorin is the last of the major proteins we will consider. This is also an intrinsic protein of the membrane and, like band III, it is a transmembrane glycoprotein. Most of the mass of glycophorin is on the external face of the bilayer where about 100 sugar residues are located in 16 sidechains. *A single hydrophobic segment penetrates the membrane, leading to a hydrophilic tail exposed to the cytoplasm.* Despite a wealth of structural data on glycophorin, little is known of its function. The carbohydrates of glycophorin determine the ABO and MN blood group antigens and one particular sugar, sialic acid, confers a high negative charge to the cell surface. This may be important in the life cycle of the erythrocyte as it has been shown that cells lose sialic acid as they age in the circulation system. Correlated with this is the observation that *loss of sialic acid is a signal for removal and destruction of an erythrocyte by the spleen and liver.* In this way the lifespan of red blood cells may be regulated.

Initial ideas about the organisation of proteins in biological membranes were developed to account for the results of surface tension measurements of membranes. Davson and Danielli originally supposed that proteins existed as

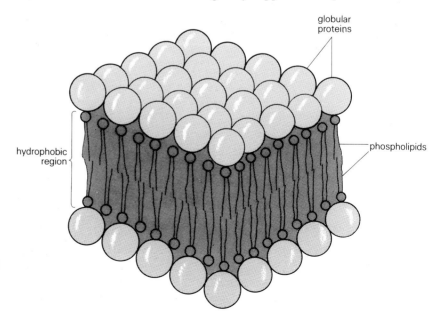

Figure 4.7 Membrane model proposed by Davson and Danielli which shows proteins as being covalently bound to the lipid molecules.

covalently bonded globular structures bound to the polar ends of lipids (Fig. 4.7). Later they developed the paucimolecular model in which the protein appears to be smeared over the hydrophilic ends of the lipid bilayer with protein-lined pores penetrating the membrane at intervals. This model remained popular for many years, being reinforced by the early structural data provided by electron micrographs. With appropriate staining these showed membranes to be made up of a layer of electron-opaque material separating two layers of electron-dense layers separated by a lighter area in between. This picture was interpreted as confirming the protein-lipid-protein model of Davson and Danielli and it was some time before this model came to be questioned. The view that the image of the membrane as revealed by conventional electron microscopy provided a true picture of the molecular organisation of membranes was first upset by developments in electron microscopy techniques. A method was developed for looking at the hydrophobic interior of membranes called freeze-fracture electron microscopy (see §1.4.4). In this procedure cells are frozen rapidly in liquid nitrogen ($-196°C$) and fractured. The plane of fracture passes through the hydrophobic interior of membranes in preference to other pathways, splitting the bilayer into two monolayers.

Membranes prepared in this way are seen to be covered with bumps and depressions about 7-8 nm in diameter which are normally randomly distributed (Fig. 4.8). Work with artificial lipid bilayers and labelled membrane proteins have confirmed that the particles seen in these membranes are proteins, the so-called *intramembrane particles*, which traverse the bilayer.

Closely following this major discovery came a second procedure developed to examine the distribution of surface proteins. Human and mouse cells have certain species-specific plasma membrane proteins which can be selectively labelled with antibodies. When a human and mouse cell are fused together to form a hybrid cell or **heterokaryon**, both the human and mouse surface proteins can be seen to be confined to their own halves of the heterokaryon. After about 30-50 minutes at 37°C, however, the surface proteins begin to intermix so that mouse and human proteins become randomly distributed over the whole surface of the hybrid cell (Fig. 4.9). This elegant experiment demonstrated that *plasma membrane proteins are mobile in the plane of the membrane*. It seems as though the fluid properties of the lipid bilayer enable membrane proteins to diffuse rapidly. So far, no evidence of a flip-flop mechanism has been seen for proteins.

4.1.3 THE FLUID MOSAIC MODEL

We are now in a position to assemble the data that are available on membrane composition and organisation into a model of membrane structure. Like all models, it does not pretend to be a definitive statement of how membranes really are, but a description of a structure which satisfactorily explains the observations made by numerous scientists over a period of many years.

The contemporary model for membranes, the **fluid mosaic model**, incorporates many of the ideas put forward by Gorter and Grendel in the 1920s and Davson and Danielli in the 1930s and 1950s. These early ideas were first formalised in 1959 as the '**unit membrane**' concept by Robertson (Fig. 4.10). Largely on the basis of electron micrographs, he suggested that all cellular membranes are built upon the structure proposed by Davson and Danielli.

Figure 4.8 Freeze-fracture
image of the P-face of a cardiac
muscle cell plasma membrane
(sarcolemma) × 96 000. The
small 'bumps' are intramembrane
particles (approximately
5–10 nm diameter), thought to
be integral proteins. The larger
pits may be the sarcolemmal
caveolae.

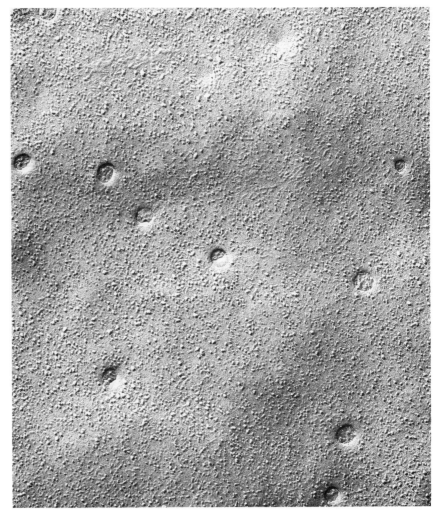

Figure 4.9 Diagrammatic
representation of the rapid
intermixing of cell surface
antigens observed by Frye and
Edidin.

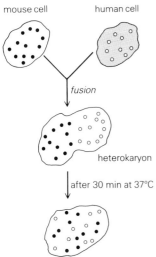

Key
• surface antigen specific to mouse cell
○ surface antigen specific to human cell

Dissatisfaction with this model of membrane structure arose with the
findings of protein mobility and intramembrane location. Neither could be
accounted for by the Robertson-Davson-Danielli model and so Singer and
Nicolson advanced a modified hypothesis. Their fluid mosaic model retains
the lipid bilayer as the framework of the membrane (Fig. 4.11), while proteins
are embedded in and float in the bilayer. According to the model, proteins can
either penetrate through the bilayer or be embedded in only one side of the
membrane. Singer and Nicolson also defined the integral and peripheral
distribution of proteins and proposed that lateral movements of membrane
proteins resulted in aggregation into functional protein assemblies.

Membrane carbohydrates, attached to glycolipids or glycoproteins, were
envisaged in the fluid mosaic model to be involved in cross-linking so as to
anchor membrane components to each other and thus restrict lateral
movements. The membrane asymmetry of carbohydrates and proteins was
also introduced; lipid asymmetry was included at a later stage as more
evidence accumulated on this phenomenon.

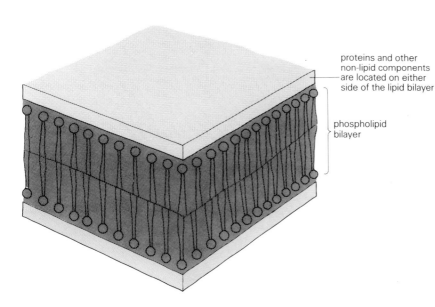

Figure 4.10 The unit membrane as described by Robertson.

proteins and other non-lipid components are located on either side of the lipid bilayer

phospholipid bilayer

The experimental evidence that has been collected since the first proposal of the fluid mosaic model has richly supported this concept of membrane organisation. Today it forms the basis of much of our understanding of the role of membranes in the cell and it seems likely that apart from minor adjustments, the model will stand for some considerable time to come.

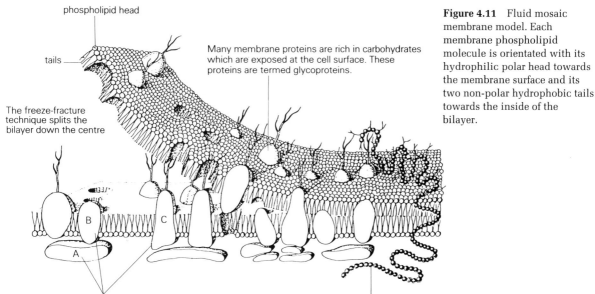

phospholipid head

tails

Many membrane proteins are rich in carbohydrates which are exposed at the cell surface. These proteins are termed glycoproteins.

The freeze-fracture technique splits the bilayer down the centre

Figure 4.11 Fluid mosaic membrane model. Each membrane phospholipid molecule is orientated with its hydrophilic polar head towards the membrane surface and its two non-polar hydrophobic tails towards the inside of the bilayer.

Membrane proteins form an integral part of the bilayer. They are located adjacent (A), embedded in one half (B), or span the entire bilayer (C). Membrane proteins also posses hydrophilic and hydrophobic regions and these are exposed at the surface or 'hidden' in the membrane in the same way as the lipids

The major protein of the red blood cell membrane is composed of approximately 200 amino acids. Twenty-five amino acids near the centre are unchanged and reside "within" the membrane. The two ends of the molecule are charged (hydrophilic) and are exposed at the surface. This protein is termed glycophorin

4.2   Membrane transport

The mechanism by which membranes transport large molecules (macro-molecules) is very different from that used in small solute and ion transport. We shall first consider how cells deal with macromolecules, processes associated with the phenomenon of *membrane fusion*.

4.2.1   EXOCYTOSIS AND ENDOCYTOSIS

Many cells secrete macromolecules which then either diffuse into the surrounding interstitial medium or become incorporated into an extracellular matrix. This process almost always involves a fusion between the plasma membrane and the membrane surrounding an intracellular vesicle which then releases its contents to the exterior. This mechanism is called **exocytosis** and is the end point of a route starting at the endoplasmic reticulum by which proteins destined for secretion are processed in a series of membrane-

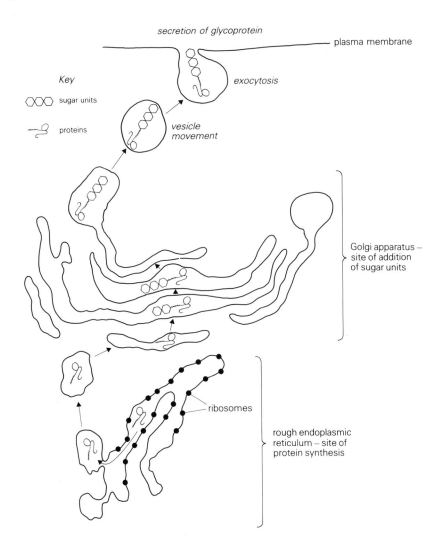

Figure 4.12 Passage of freshly synthesised proteins through the cell prior to secretion.

associated events. Secretory proteins are synthesised by ribosomes (see §6.8) attached to the rough endoplasmic reticulum (RER). During their assembly the proteins are transferred to the lumen of the RER and are later transported to the Golgi apparatus via vesicles which have pinched off from the endoplasmic reticulum. In the Golgi the proteins are combined with appropriate carbohydrate residues to form glycoproteins and packaged into further vesicles that bud off from the Golgi (Fig. 4.12) and migrate to the plasma membrane where they fuse and exocytose (Fig. 4.13).

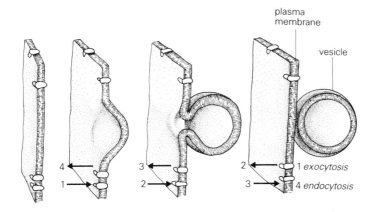

Figure 4.13 Fusion between membranes of the cell surface and carrier vesicles in exocytosis. Endocytosis requires membrane fission (or budding) to form vesicles.

The secretory vesicle membrane which becomes *incorporated into the plasma membrane* is later retrieved by the *counterprocess of endocytosis*. All eukaryotic cells continually produce pinocytotic vesicles; these small endocytotic entities trap bathing medium within them during endocytosis and this is also ingested. The rate of this so-called '**fluid-phase endocytosis**' varies according to cell type, but whatever the rate, *a large proportion of the plasma membrane is internalised by this process*. The fact that cells retain their surface area and volume is explained by the replacement of internalised membrane by exocytotic vesicle fusion with the plasma membrane. We know that the rate of biosynthesis and degradation of membrane is very low in comparison with the replacement rates measured for the plasma membrane, so it is assumed that the bulk of the endocytosed membrane is recycled by a mechanism which is not yet understood.

Most endocytotic vesicles enlarge by mutual adhesion and fusion and then fuse with the lysosomes, organelles of intracellular digestion. In some cells, specialised endocytotic vesicles avoid the lysosome system altogether and are able to transfer their contents to other sites; usually another region of the plasma membrane. In this way, *molecules can be transferred from one side of a tissue layer to another*.

Fluid-phase endocytosis is not the only pathway through which macromolecules enter cells. In most animal cells a more specialised mechanism exists called '**receptor-mediated endocytosis**' in which certain molecules present in the bathing medium are selectively internalised. The great advantage of receptor-mediated endocytosis lies in the fact that certain molecules are *selectively concentrated in the endocytotic vesicle thereby reducing the needless ingestion of large volumes of extracellular medium*. The vesicles associated with receptor-mediated endocytosis are found to be coated with distinctive proteins on their cytoplasmic face. The best

Figure 4.14 Receptor-mediated endocytosis and the role of the coated pit.

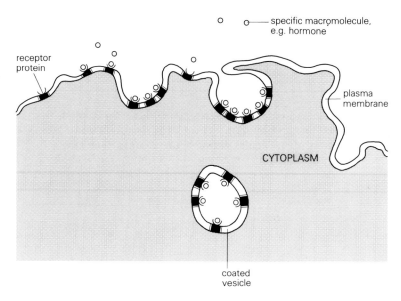

characterised protein to date is clathrin, a fibrous protein of 180 kdaltons, which complexes with other smaller proteins to form a polyhedral coating on the surface of the **coated vesicle** (Fig. 4.14). Coated vesicles are thought to originate as **coated pits**, small regions of the plasma membrane which act as sites for receptor-mediated endocytosis. In this process, macromolecules present in the medium bind to specific membrane receptors which are integral proteins. These receptors may be distributed over the whole surface of the cell or be restricted to the sites of coated pits. In the former case, bound receptor complexes migrate in the plane of the bilayer until they become associated with a coated pit prior to internalisation to form a coated vesicle.

4.2.2 DIFFUSION

Above absolute zero, molecules are in constant motion. The energy of motion, called *kinetic energy*, confers a chemical potential to the mass of molecules contained in a certain compartment. The greater the concentration of molecules in a compartment, the greater the kinetic energy and chemical potential of the compartment. When two compartments of differing molecular concentration are allowed to associate, the difference in chemical potential between compartments will generate a net movement of molecules down the chemical potential gradient until the molecules are evenly distributed and the chemical potential in all parts of the system is the same. This net movement is called **diffusion**, and the introduction of a membrane between the two compartments will have no effect on the system provided that all of the molecules can pass through the membrane with equal ease.

Diffusion as described above is greatly complicated in cases where a semipermeable or selectively permeable membrane separates the two compartments. All biological membranes have this property as they *allow some molecules or ions to pass through much more readily than others*. Glucose, for example, passes into cells at a rate some 100 000 greater than that predicted from a knowledge of the solubility of this sugar in lipid, the major constituent of membranes. As the energy required for glucose transport is provided by a gradient in chemical potential (i.e. diffusion down a concentration gradient) rather than from a source of cellular energy, the transfer of glucose in this example must be considered as a passive process. Such systems are called **facilitated transport**.

4.2.3 FACILITATED TRANSPORT

The rate of movement of molecules across membranes by facilitated transport is greatly affected by concentration. As we have seen, in simple diffusion the rate of penetration of molecules through membranes increases in a linear fashion with increasing concentration. In contrast, facilitated movement *shows saturation effects, in that the linear relationship found at very low concentrations rapidly drops off at higher concentrations to reach a maximum rate* (Fig. 4.15).

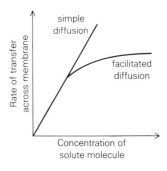

Figure 4.15 Comparison between the kinetics of simple diffusion and facilitated diffusion.

This behaviour has similarities with that of enzymes in catalytic reactions. In this case, as substrate concentration rises, the enzymes become saturated and the rate of reaction begins to level off. The similarity in behaviour between the two systems has given rise to the belief that facilitated diffusion is *carried out by protein carrier molecules embedded in the membrane*. Other features of facilitated diffusion serve to emphasise the parallels with enzyme reaction kinetics. *There is a high degree of specificity*; each molecule to be transported is carried by a separate carrier specific to that molecular species. Inhibition of the rate of transport is seen when molecules that are closely related to the substance normally transported are allowed to compete for the binding site on the carrier (compare competitive inhibition, §3.5.2). The proteins responsible for facilitated transport can be envisaged as integral membrane proteins with specific binding sites exposed on the aqueous phase. Under appropriate conditions, these binding sites will become occupied by the ion or small molecule to be transported and it will then be carried to the opposite face of the membrane and released.

The way in which carrier proteins shuttle their substrate molecules across membranes is not clearly understood. It is possible that the protein acts as a mobile carrier, but the energy required to accomplish this feat, in which polar regions of the protein would have to pass through the hydrophobic interior of the membrane, would be very high indeed. Another possibility, which seems more compatible with the fluid mosaic model, is that carrier molecules are transmembrane proteins. The shape of the protein would allow for the formation of a channel or pore through the centre of the molecule once binding of the substrate has occurred. Following the transfer of the substrate molecule through the channel, the carrier protein would return to its previous state ready to accept another molecule (Fig. 4.16).

The very limited experimental evidence on facilitated transport supports the channel model. It seems that a *change in the configuration of the carrier protein is induced by the occupation of the receptor site by substrate*. The folding of the amino acid chain creates a transient pore in the sense that the binding site is reorientated to the opposite face of the membrane, and the substrate dissociates from the carrier to go into solution. The amino acid chain then returns to its original configuration with the channel closed.

The mechanism of facilitated transport accounts for the passage of a wide variety of substances across membranes. Since both facilitated and passive diffusions are driven solely by potential gradients (concentration differences) and *require no added energy input from the cell*, they can be regarded as very

Figure 4.16   Facilitated
diffusion by (a) mobile carriers
or (b) channel-forming carrier
molecules. In (a) the
concentration gradient drives
the release of the transported
molecule from its specific
binding site (3) and the carrier
will return to its original
location (1). In (b) there is a
specific binding site on the
carrier which regulates which
molecules can pass through the
channel.

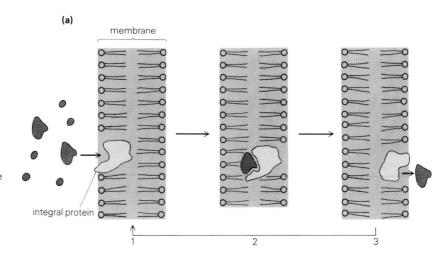

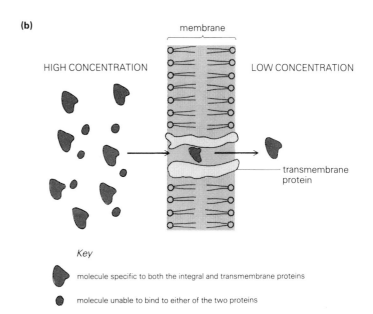

efficient mechanisms. There is a clear disadvantage associated with these
processes, however: if a cell requires molecules to be present at concentrations greater than those found in the surrounding medium, they cannot enter
the cell by passive or facilitated diffusion. In such circumstances, **active
transport** is required.

4.2.4   ACTIVE TRANSPORT

Many transport processes, such as the uptake of nutrients by intestinal
epithelial cells, work against a concentration gradient; such mechanisms
require the input of energy. In very many respects, active transport systems
closely resemble facilitated diffusion processes: binding specificity, saturation
kinetics and competitive inhibition are all features of active transport

mechanisms. In its most simple form, active transport can be imagined as a facilitated diffusion system linked to a supply of cellular energy, i.e. there is a membrane protein acting as a carrier for some specific molecular species which works in a manner very similar to that described for the facilitated diffusion mechanism. The fundamental difference is that *active transport systems are able to transport materials against a concentration gradient*, thus providing a means for cells to obtain considerable quantities of certain metabolites even when these are present at very low levels in the surrounding environment.

There is *no single mechanism for the coupling of transport systems to a supply of energy*, but there are basic concepts which are fundamental to all active transport mechanisms across biological membranes. A variety of energy sources are used to power the uptake of solutes against a concentration gradient. ATP is widely used, as are proton gradients (see ★2.4.1) and the analogous sodium-dependent systems and examples of these will be described here. It is important to remember, however, that more complex processes also exist using light (chloroplasts) or oxidative metabolism (inner mitochondrial membrane) as the primary energy source.

4.2.5 ATP-LINKED SYSTEMS

In this type of mechanism, the cytoplasmic face of a carrier protein contains a binding site for ATP (Fig. 4.17). When a molecule of ATP in the cytoplasm collides with an exposed binding site of the protein, it will attach and will be converted to ADP with the removal of a phosphate. The consequent energy release causes a change in the conformation of the carrier protein such that a binding site for the substance being transported becomes exposed on the outer membrane face. The change in the folding pattern of the carrier protein is energetically unfavourable, and as soon as the external binding site has become occupied by the specific molecule to be transported, *the carrier is free to return to an energetically more stable shape*. In so doing, the transported molecule is drawn through a hydrophobic channel in the carrier protein to the cytoplasmic face of the membrane. The change of conformation of the carrier now renders the transported molecule's binding site into a low affinity shape, and the transported substance is released into the cytoplasm. Release of the phosphate group accompanies this event, triggering the final change in protein shape which returns the carrier to the original state in which it is once again ready to accept ATP.

Active transport of Na^+ and K^+ ions are amongst the most important transport mechanisms of cells. Almost all vertebrate cells and most invertebrate cells *maintain internal concentrations of these ions which are very different from their environmental values*. For most mammalian cells, K^+ concentration is 120-160 mmol inside, compared with less than 4 mmol outside; the external Na^+ concentration is about 150 mmol while the value inside the cell is more like 10 mmol. The plasma membranes of nearly all animal cells contain an Na^+-K^+ pump that operates as a so-called **antiport** (Fig. 4.17); a carrier protein that actively pumps Na^+ out of the cell while actively pumping in K^+ at the same time. It has been estimated that *more than a third of a cell's energy requirement is used up by this pump alone* as it works to maintain the concentration differences across the membrane. It was found in 1957 that an enzyme that hydrolyses ATP to ADP requires Na^+ and K^+ for optimal activity. This ATPase was also shown to be inhibited by the

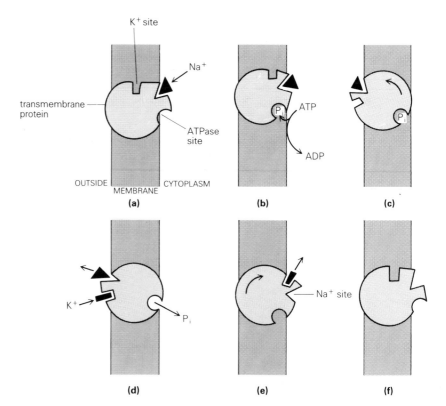

Figure 4.17 A model system for Na$^+$, K$^+$–ATPase active transport system: (a) transmembrane protein with three specific sites, Na$^+$ bound to the receptor site; (b) binding of ATP and conversion to ADP + P; (c) change in protein conformation; (d) release of Na$^+$ into the external medium and binding of K$^+$ at a different site; (e) transmembrane protein returns to its most favourable energetic shape; (f) release of K$^+$ into the cytoplasm.

drug **ouabain** which was known as an inhibitor of the Na$^+$-K$^+$ pump; it was eventually found that the transport of Na$^+$ and K$^+$ is closely coupled to ATP hydrolysis. For every molecule of ATP hydrolysed, Na$^+$ ions are pumped out and K$^+$ ions are pumped in. Subsequent work isolated the polypeptides of the pump from plasma membranes. The whole complex, with a total molecular weight of about 250 kdaltons, spans the membrane, has carbohydrate residues on it like many other membrane proteins and has binding sites on the inner-facing portion for Na$^+$, ATP and Mg^{2+} with the K$^+$ binding site directed towards the outside.

These findings have been used to construct a model of the Na$^+$, K$^+$-ATPase active transport system. The carrier complex spans the plasma membrane and contains a hydrophilic inner channel. To begin with, the complex binds Na$^+$ at the cytoplasmic face of the membrane. This activates the ATPase activity of the complex; as ATP binds to the carrier it is hydrolysed to ADP and the phosphate group is transferred to the complex (Mg^{2+} is required for this step). As in our general model, binding of the phosphate group changes the conformation of the carrier such that the Na$^+$-site is transferred to the outside of the membrane. The same movement also exposes the K$^+$-binding site which binds the ion. This reaction leads to release of Na$^+$ to the outside, and the phosphate group is also released at this time from its binding site on the inner face. The complex now returns to the original conformation, carrying the bound K$^+$ to the inside of the membrane. On completion of this change of shape, the K$^+$-binding site loses its affinity for this ion and the K$^+$ is released to the cytoplasm. The carrier is then free to resume Na$^+$ transport if required.

In fact, one Na$^+$ is not exchanged with one K$^+$ in this way; instead the complex works by binding three Na$^+$ and two K$^+$ ions with each turn of the cycle. It is also important to realise *that only a very small segment of the carrier complex actually moves from one side of the membrane to the other; the bulk of the carrier remains in a stable state in the membrane.*

4.2.6 Na$^+$ SYMPORT OF SUGARS AND AMINO ACIDS

Active transport of Na$^+$ and K$^+$ is of special significance to animal cells because the movement of these ions across the plasma membrane generates an electrical potential across the membrane. It is this potential which is also called the resting potential of nerve cells. The Na$^+$, K$^+$-ATPase system also establishes the Na$^+$ gradient that is used to drive the active transport of sugars and amino acids into cells. The Na$^+$ that is pumped out of cells will tend to move down a concentration gradient back into the cell. Generally, the plasma membrane is impermeable to the ion except at sites where carrier proteins

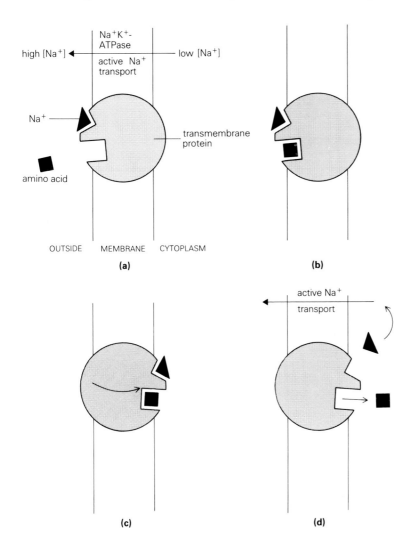

Figure 4.18 Model of amino acid or sugar/Na$^+$ symport transport system.

exist to act as facilitated diffusion transporters. However, the carrier protein in this case will have a *second binding site for a specific sugar or amino acid which must be occupied before Na$^+$ can cross the membrane by facilitated diffusion.* As the carrier protein transfers the Na$^+$, it also shifts the bound sugar or amino acid to the cytoplasmic face of the membrane. The sugar or amino acid will therefore be moved even when there is a concentration gradient acting *against* it, as it is the Na$^+$ concentration gradient which provides the energy to relocate the carrier protein binding sites to the cytoplasmic side of the membrane (Fig. 4.18). Active transport of this type is called either **cotransport** or more frequently, **a symport** system.

Na$^+$ symport systems of active transport are common. In mammals, for instance, they are the basis for the transfer of sugars and amino acids from the lumen of the gut into the epithelial cells which line the intestine.

4.3   Cell junctions

The differentiated cells of multicellular organisms are normally organised into tissues and organs, co-operating with each other in their various metabolic functions. Many such tissue cells are linked to each other by areas of specialised plasma membrane collectively termed **cell junctions**. There are three categories of animal cell junction (Fig. 4.19) and one plant cell junction (Fig. 4.20).

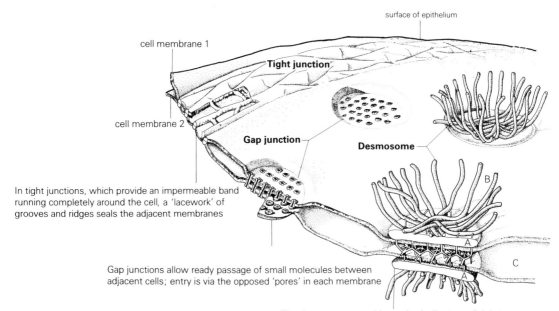

In tight junctions, which provide an impermeable band running completely around the cell, a 'lacework' of grooves and ridges seals the adjacent membranes

Gap junctions allow ready passage of small molecules between adjacent cells; entry is via the opposed 'pores' in each membrane

The desmosome provides a physically strong link between adjacent cell membranes. On the cytoplasmic sides of the desmosome, dense plaques (A) are anchored by numerous intermediate (desmin) filaments (B). Between the cells, adjacent membranes are linked by a carbohydrate-rich material (C). Desmosomes may also occur in a band form, joining large areas of adjacent membrane

Figure 4.19   Cell junctions in animal cells.

4.3.1 DESMOSOMES

Desmosomes are common in animal tissues and are thought to be involved in structural support, especially in tissues such as heart muscle and the skin epithelium. They do not appear to be involved in cellular transport mechanisms and generally can be seen in electron micrographs to be closely associated with intracellular filaments. In thin section, the parallel plasma membranes of two adjacent cells come to lie some 15-30 nm apart (depending on the tissue) with an electron-dense material filling the gap between the membranes. This material is now thought to be of filamentous nature and serves to hold the interacting membranes together. In epithelial cells, **tonofilaments** (made of keratin) extend from one side of a cell to another and are anchored in the plasma membrane at the sites of desmosomes. Together with the intercellular filaments, the tonofilaments act as a structural support stretching from cell to cell in the epithelial tissues of the body.

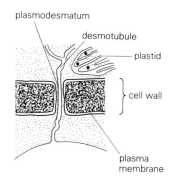

Figure 4.20 Plants use plasmodesmata to link cells into communicating groups of individuals.

4.3.2 TIGHT JUNCTIONS

Tight junctions appear as a continuous belt of contact between epithelial cells. They serve as highly selective permeability barriers, maintaining the barrier functions of cell sheets by isolating two fluid phases that can have very different chemical compositions. In thin section, it is seen that the intercellular space between plasma membranes is lost. Tight junctions are formed when two interacting plasma membranes make direct physical contact. Freeze-cleaved replicas expose tight junctions as linear rows of intramembrane particles, sometimes interdigitating in complex furrows and ridges across the width of the encircling belt of cell contact (Fig. 4.19).

Invertebrates lack tight junctions, but the vital barrier role of epithelial cell sheets is maintained by **septate junctions**; here the interacting plasma membranes are joined across an intercellular space by parallel rows of proteins which themselves form a seal. The regular periodicity of these junctional proteins serves to give a ladder-like appearance to septate junctions when viewed in thin section.

4.3.3 GAP JUNCTIONS

Gap junctions are the most common of the animal cell junctions. They are primarily concerned with cell to cell communication as it was shown some time ago that small molecules and ions of the cytosol could readily move from cell to cell via such junctions. It has been shown for cultured cells that molecules such as nucleotides, sugars, vitamins and amino acids can readily cross between coupled cells. Large macromolecules such as nucleic acids and proteins are excluded, suggesting a functional pore size of about 1.5 nm.

In thin section, the gap junction appears as an area where the participating plasma membranes come to lie about 2-3 nm apart. The junctional areas are composed of particles some 7-8 nm in diameter, often arranged in hexagonal arrays consisting of more than 100 particles, though smaller gap junctional complexes are also common (Fig. 4.19). High-resolution studies indicate that each 7 nm particle is composed of six protein sub-units arranged so as to form a cylinder about a central hydrophilic channel. Gap junction formation is very rapid and does not require new protein synthesis by cells. This evidence suggests that junction formation *is a self-assembly mechanism induced by cell contact and utilising specific gap junctional proteins which have been randomly distributed in the plasma membrane.* It is assumed that the particles of the two contacting surfaces come to lie in register, leading to the formation of an open aqueous channel between the adjacent cells.

4.3.4 PLASMODESMATA

The cell wall of plant cells would seem to act as a barrier to any form of cell to cell junction. Intercellular communication does occur in plant tissue, however, mediated by the **plasmodesmata** (Fig. 4.20). Apart from a few cell types, plant cells are connected to each other by fine cytoplasmic channels which pass through the cell walls. A plasmodesmatum is a cylindrical, membrane-lined channel with a diameter between 20 and 40 nm. Within this channel, most plasmodesmata contain a narrow cylindrical body, the **desmotubule**, which can be seen in electron micrographs to link the endoplasmic reticulum of the connected cells. The use of dyes has produced evidence that plasmodesmata allow the passage of small molecules and ions from one cell to another very much in line with the gap junctions of animal cells. Similarly, despite their diameter, free exchange of macromolecules cannot be demonstrated, so some selectivity and control over what can cross the junction must be invoked for plasmodesmata. Certain plant viruses can overcome the block to transit of macromolecules, providing evidence that in normal circumstances *this block must be a physiological rather than physical process.* In the same way as animal cell gap junctions, plasmodesmata allow *individual cells to share ions and small metabolites which serves to confer a group identity to what may otherwise be rather isolated cells.*

4.4 Cellular recognition and adhesion

4.4.1 GLYCOPROTEINS AND GLYCOLIPIDS

As we have seen, animal cell plasma membranes contain glycoproteins and glycolipids which extend their carbohydrate-bearing portions into the extracellular spaces. There has been much interest in the role of such carbohydrates, and the asymmetric distribution of these molecules has prompted the idea that these components of the plasma membrane are responsible for the many aspects of cellular recognition. We shall restrict our examination of this function of glycoproteins and glycolipids to animal cells as the surfaces of plants, fungi and bacteria are largely in contact with a cell wall which greatly hinders proper analysis in these organisms.

The sugar complexes of glycoproteins and glycolipids are composed of approximately six monosaccharides which are capable of being combined in a multitude of variations. A large body of evidence has accumulated over the last decade which indicates that the *various patterns of oligosaccharide complexes form the basis of cellular recognition and adhesion phenomena.* Some of the examples described below are more fully understood than others, and in this field of research very much more work is needed before complete descriptions can be given with authority.

4.4.2 RECEPTOR FUNCTION

Membrane glycoproteins (and glycolipids) can act as surface receptors for a range of circulating agents. One important example of this is the capacity of certain **gangliosides** (glycosphingolipids) to act as receptors for hormones of protein nature (as distinct form the steroid hormones). Binding of hormones such as **luteinizing hormone** (LH) and **thyroid stimulating hormone** (TSH) to their target cells can be blocked by specific gangliosides which bind avidly to the hormone and thus block interaction with the membrane receptor. Receptors for other protein hormones are known to consist of glycoproteins, though in these cases the molecular basis for the binding specificity is very poorly understood. Whatever the details, the cell receptor-hormone inter-action acts as a signal to change the behaviour of the target cell. Experiments using radioactively labelled protein hormones have shown that in most cases the surface receptor-hormone complex enters the target cell by receptor-mediated endocytosis. The hormone does not act as the intracellular signal, however, as it is normal for endocytotic vesicles to be carried to the lysosomes where degradation of the hormones is executed. Evidence supporting the belief that it is the binding of hormone to a specific surface receptor that stimulates the target cell comes from work on TSH and insulin receptors. Target cells for these two hormones will respond even when the surface receptor is bound not to the hormone, but to an antibody specifically made to bind to the appropriate receptor molecule (Fig. 4.21).

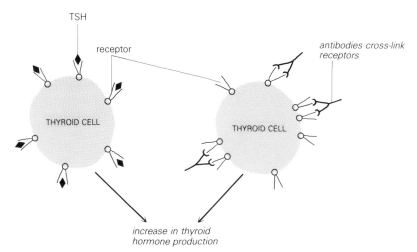

Figure 4.21 Cell-surface receptors for thyroid stimulating hormone (TSH) can stimulate the correct physiological response in target cells even if they are bound to antibodies rather than TSH.

Receptors for viruses and bacteria also exist at the cell surface. Viruses proliferate inside cells and therefore need first to attach to cells and then to be engulfed. Most of our knowledge in this area comes from studies on organisms such as the influenza virus which is one of a group of **enveloped RNA viruses**. The envelope of such viruses arises as a consequence of the budding of the mature virus out of an infected cell. Part of the plasma membrane of the host cell coats the virus particles as they bud off, and there is evidence to show that while the lipid composition of the viral envelope is similar to that of the host cell plasma membrane, *the protein content is quite different*. It seems that proteins coded for by the viral genome become inserted at specific sites of the host surface and it is at these sites that budding can occur. As an experimental model, the binding of influenza virus particles to erythrocytes has been extensively studied. Viral envelope glycoproteins bind specifically to receptor sites on the erythrocyte membrane now known to be the glycoprotein, glycophorin. The binding is mediated by one type of sugar residue called sialic or neuraminic acid present on glycophorin; destruction of this sugar by the enzyme neuraminidase will abolish virus attachment to the erythrocyte.

Less convincing data exist to suggest that certain bacteria also recognise and bind to sugar-containing target sites on the surfaces of mammalian cells. Some strains of the common gut bacillus, *Escherischia coli*, bind specifically to D-mannose residues on human epithelial cells and a mannose-specific protein has been isolated from the bacterium. Similar findings exist for the bacterium *Salmonella typhi*, while the sugar fucose is thought to be the receptor molecule for *Vibrio cholera*.

4.4.3   BLOOD GROUP AND HISTOCOMPATIBILITY ANTIGENS

The examples we have discussed in this section so far are cases of membrane receptor behaviour. Surface carbohydrate is also involved in activities more accurately termed **recognition behaviour**, by which is meant systems developed to receive information from the environment in which a cell finds itself. The ability to respond to the information in some way usually resides in the cell. A classic example of cell recognition is the ABO blood grouping first discovered by Landsteiner in 1900. This blood group classification is based upon the occurrence of **antigens** on human erythrocytes, epithelial and endothelial cells. These surface antigens have predictable relationships to groups of circulating antibodies in the serum which are glycoprotein in nature, whereas the surface antigens are now known to be carried by both glycoproteins and glycolipids of the plasma membrane. It is thought that *the ABO specificity is carried by individual specific monosaccharide units carried on a backbone of a common precursor oligosaccharide.* Each of the four genes involved in expression of the ABO antigens *produces an enzyme that attaches the relevant specific sugar to the precursor oligosaccharide.*

The ABO blood group antigens are actually members of a larger group of surface components called the **histocompatibility antigens complex** (HLA complex). These are cell-surface antigens that are specific for the individual and responsible for the phenomenon of tissue rejection in transplants. In man, the HLAs, or **transplantation antigens**, are coded for by genes on chromosome 6 and are thought to consist of glycoproteins of about 44 000 molecular weight which are located as transmembrane elements of the plasma membrane. The *antigenic specificity of this group of glycoproteins seems to reside in the detailed sequence of the amino acid chain rather than in the carbohydrate sidechain.*

4.4.4   CELL ADHESION

Cell adhesion is a cell to cell recognition phenomenon of *crucial importance in the formation of tissues and organs during embryonic life.* During embryogenesis, the cells of an embryo undergo a series of co-ordinated events which involves the selective adhesion between cells later destined to differentiate into the major tissues and organs of the body. The sorting out of embryonic cells is thought to depend upon *specific intercellular recognition processes which are responsible for the selective adhesive events* so vital to normal development. Experiments using well differentiated tissues such as heart, liver, brain and many more have shown that the adhesive specificity is *retained by differentiated, post-embryonic cells*. Cells that have been separated from various tissues will preferentially reassociate with cells of the same tissue when allowed to mix with cells from other sources. Cells obtained from liver will adhere together in preference to mixing with heart cells, for instance. It has long been thought that this type of tissue specificity must reside in the glycoprotein components of the plasma membrane. This is the only cell organelle which mediates cell to cell contact when adhesions occur between cells, and the glycoproteins, with their varied amino acid and carbohydrate composition, *are ideal candidates for the specific binding between complementary adhesive molecules*. Despite the considerable efforts of many groups of scientists, however, no completely satisfactory explanation of adhesive recognition and adhesion has yet been proposed. Recently a model based upon a binding between a carbohydrate group and a carbohydrate-binding protein has become popular, largely because such proteins have frequently been found on the surface of cells (Fig. 4.22). The carbohydrate-protein interaction linking one cell with another may be direct, in that both molecules are integral components of the plasma membrane, or may be mediated by some secreted aggregation factor released into the intercellular space at some critical period of embryonic development. Although several examples of cell-surface interactions of this nature have now been discovered in organisms as diverse as sponges and man, it is too early to say whether such mechanisms are really used for generating the selective cellular associations within embryonic tissues.

Before leaving this section, mention must be made of cellular adhesion not to other cells, but to a substrate of secreted molecules found deposited as a matrix between cells. This extracellular matrix, which varies in composition according to the type of cell producing it, will be discussed in more detail in Chapter 7. Cell adhesion to the matrix is known to be mediated by specific binding sites for the cell surface found on certain molecules of the matrix. One such molecule called fibronectin has been exhaustively studied over the last decade. It is a glycoprotein, secreted by fibroblasts in the main, which cross-links cell surfaces (Figs 4.23a & b) with the major matrix protein,

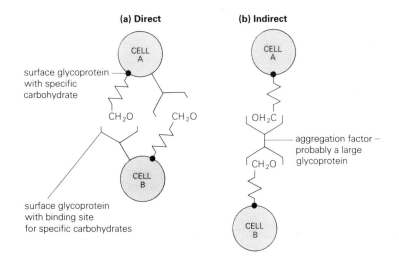

(a) Direct

CELL
A

surface glycoprotein
with specific
carbohydrate

CH$_2$O   CH$_2$O

CELL
B

surface glycoprotein
with binding site
for specific carbohydrates

(b) Indirect

CELL
A

OH$_2$C

CH$_2$O

CELL
B

aggregation factor –
probably a large
glycoprotein

Figure 4.22 Models of cell–cell adhesion based on protein–carbohydrate interactions.

Figure 4.23 The promotion of cell spreading by fibronectin. Human skin fibroblasts seeded onto (a) glass or (b) fibronectin-coated glass were allowed to spread for 100 minutes. Both prints × 915.

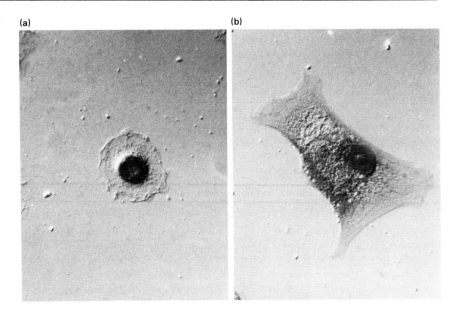

collagen. Since the *extracellular matrix plays a major role in controlling cell proliferation and the organisation of tissue masses,* much work has been done on the specificity of cell-substrate adhesion. It has been shown, for instance, that while fibronectin can bind fibroblastic cells to collagen, it is not effective in binding epithelial cells. Another glycoprotein, laminin, is required in this case and is the fibronectin analogue for cells of epithelial nature. Other such specific matrix attachment factors are being discovered, all of which seem to promote the matrix adhesion of a limited group of cell types. This type of *specificity could be important in the differentiation of complex organs consisting of several tissue types such as muscle and cartilage.*

4.5 Trends in research

Membranes have a diversity of functions, ranging from the protective role of the plasma membrane, through the compartmentalisation of cellular reactions and transport of substrates, to their roles in the maintenance of electro-chemical gradients and recognition processes. All of these areas are the source of considerable research at the present time and cannot be covered comprehensively here. A few topics which serve to indicate the breadth of present research on membranes have been chosen.

4.5.1 PHOTOCHEMICAL REACTIONS OF MEMBRANES

Visible light is a form of energy that can be thought to travel in waves with lengths between about 390 and 760 nm. The wave path is not radiated as a continuous beam but exists as discrete packets of energy known as **photons** or **quanta**. Certain molecules can absorb the energy of photons, the most commonly known biological molecule with this attribute being the photosynthetic pigment chlorophyll. This pigment absorbs red and blue light (and therefore appears green) with the result that electrons occupying certain orbitals in the molecule are raised to new orbitals with higher energy levels. This state (the excited state) is unstable and, in the case of chlorophyll, the *orbital occupied by the excited electron extends far enough so as to overlap vacant orbitals in a second molecule.* Transfer of an excited electron to a stable orbital in the second molecule may then occur. Once transferred, the energy of the electron *has been captured as chemical energy.* The light energy absorbed by the chlorophyll in photosynthesis is converted to chemical energy in this way.

The complexities of photosynthesis will be covered elsewhere in this series. Green plants also contain another photosensitive molecule which is currently the subject of intense research interest. The molecule in question is **phytochrome**, which has been known for some time to exist in two forms. The type known as Pr absorbs most light at around 660 nm and is converted by light into the second form, Pfr, which absorbs most light at about 730 nm and is converted back to Pr. The fundamental function of phytochrome is to detect the relative amounts of red and far-red light in the incident illumination of plants, via which *phytochrome controls physiological processes such as seed germination, growth of hypocotyls and stems, and leaf development.*

Recently, it has been demonstrated that phytochrome also controls more rapid or short-term responses in plants such as the *nyctinastic* leaflet movement in legumes. These movements occur in a matter of minutes and have led to the idea that phytochrome may act as a controller of *membrane permeability* in plants. Evidence for rapidly reversible red/far-red responses are now abundant and include measurements of surface membrane potential changes, transmembrane potential changes and ion flux responses sometimes occurring within seconds. All these data suggest a membrane-associated site for phytochrome action and we shall examine one particular example in some detail which provides very strong evidence for a link between phytochrome and plant cell membranes.

There are many organisms which display reversible chloroplast orientation to light intensity and direction. In low-intensity light, chloroplasts are found mainly facing the light, while at high intensity, the chloroplasts move to the flanks of the cell (Fig. 4.24). The question is, how is light perception transformed into chloroplast relocation in plant cells? In cases where phytochrome is known to be the photoreceptor pigment, it has been demonstrated by combined ultrastructural and inhibitor studies that light, via phytochrome, stimulates the uptake of calcium ions. This in turn activates the calcium-dependent actin microfilament-myosin system to be found at the periphery of the cell (see §7.2.5) which is utilised to relocate chloroplasts. It seems likely that calmodulin plays the *same controlling role here as it does in the non-muscle cells of animals.* Experimental evidence has been produced to demonstrate that calcium uptake is increased by red light, and that this is reversible by far-red light, and it has also been shown that Pfr has a high affinity for model membranes. This supports the hypothesis that *phytochrome develops a high affinity for membranes following activation by red light,* one consequence of which is a change in the calcium permeability of membranes.

In animals, the perception of light by photoreceptive molecules is now well understood. The photoreceptors of the vertebrate eye consist of cells called **cones** which require bright light and confer colour vision, and **rods** which work at low light intensities but only give black and white (monochromatic)

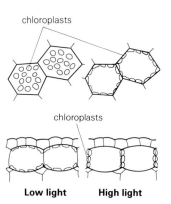

Figure 4.24 Chloroplast arrangement in the mesophyll cells of *Lemna* as seen from above and in cross section.

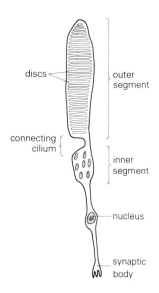

discs

outer
segment

connecting
cilium

inner
segment

nucleus

synaptic
body

Figure 4.25 Diagrammatic
representation of a retinal rod
cell.

vision. Both work in similar ways and the rod cell (Fig. 4.25) will serve as our example. The so-called outer segment of the rod cell is a cylindrical stack of about 1000 discs enclosed by the plasma membrane. Each disc is a membrane-bound organelle in which molecules of **rhodopsin**, the photosensitive molecule, are embedded. Rhodopsin is itself made up of the transmembrane protein opsin, and a smaller molecule that actually absorbs the light, 11-*cis*-retinal. Upon absorbing a photon, this molecule changes its conformation to the *trans* form and separates from opsin. This isomerisation alters the geometry of retinal (Fig. 4.26), so that in essence, a photon has been converted into atomic motion. The plasma membrane of the outer segment then becomes transiently hyperpolarised; in the dark, the plasma membrane of the rod outer segment is permeable to Na^+ but the Na^+ channels are blocked by light. Consequently, there is a decrease in the net inward flow of Na^+, resulting in hyperpolarisation. As the rhodopsin molecule that absorbs a photon is located in the disc membrane, several micrometres away from the sodium channels of the plasma membrane, there must be a mechanism by which a diffusible signal links the photolysed rhodopsin in the disc membranes to the plasma membrane. The identity of the transmitter is still a mystery, *but two candidates have recently emerged: calcium ions and cyclic GMP.* A considerable body of experimental evidence shows that sodium channels in the plasma membrane close with increasing levels of Ca^{2+} and when levels of cyclic GMP are lowered. For both calcium and cyclic GMP the converse also holds. It is also known that Ca^{2+} is removed from the rod outer segment following a light pulse; light also activates an enzyme (a **phosphodiesterase**) that hydrolyses cyclic GMP. Much more work on this problem is required before the true nature of the transmitter can safely be identified.

4.5.2 DISORDERS OF THE CELL SURFACE

Potentially, many diseases that are not clearly of viral and bacterial origin may have their basis in defects of the structure or functioning of the cell

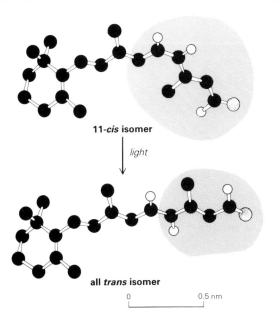

11-*cis* isomer

light

all *trans* isomer

0   0.5 nm

Figure 4.26 Primary events in the detection of light in the rod. The photon is effective in converting the 11-*cis* isomer of retinal into the *trans* form.

surface. One such example that we shall examine is an **auto-immune disease**.

The most thoroughly studied auto-immune disease is one called myasthenia gravis, in which the patient produces antibodies directed against the nerve-muscle synapse. Clinical effects of this process include impairment of transmission at the nerve-muscle junction and a resulting deterioration of muscle function. It was not until the mid-1970s that an explanation was found for the disease, which led to the development of strategies to control the worst effects.

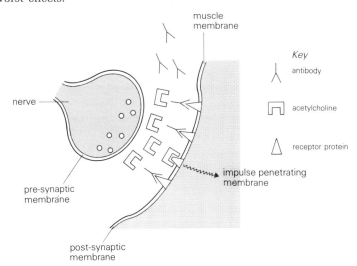

Figure 4.27 The antibodies present in patients with myasthenia gravis block the receptor protein on muscle cell membranes and prevent acetylcholine molecules from binding: thus the signal for muscle contraction is not received.

Synaptic transmission at the junction of a nerve and skeletal muscle is effected through the release of acetylcholine into the synaptic cleft. This chemical transmitter will bind at the post-synaptic membrane to specific acetylcholine receptor proteins embedded in the muscle plasma membrane. Recent work on the receptor protein has been very fruitful, and a considerable amount is known about the molecular events of receptor binding and the subsequent events leading to muscle contraction. It appears that the acetylcholine receptor protein *is the target or antigen for the auto-antibodies that have been shown to be produced in myasthenic patients*. These antibodies bind to the receptor protein so that it can no longer bind any acetylcholine present in the synaptic cleft of the neuromuscular junction (Fig. 4.27). Without this binding, muscle responses to nervous stimulation fails, leading to the clinical symptoms already described. Many laboratories are working to obtain the detailed knowledge of the receptor molecule that will be required before a totally satisfactory cure can be offered for this disease.

4.5.3   MEMBRANE TRAFFIC

The term 'membrane traffic' is used to describe the continual shuttling of membrane-bound vesicles between the various organelles of the cell cytoplasm. Considering that the role of all this vesicle traffic is to deliver specific macromolecules to their correct cellular location, it is clear that sophisticated guiding mechanisms must exist in the cell to control vesicular traffic. Very little indeed is known about how vesicles are directed where to go and this problem has received more attention recently as our knowledge of membrane organisation increases.

Unfortunately, almost nothing is yet known about the nature of those elements on the cytoplasmic face of the vesicle membrane which are presumed to be responsible for guiding the vesicles. One clue has come from work on coated vesicles which, as we have already seen, are distinguished by the clathrin coat which lines these vesicles. It has recently been shown that other accessory proteins are associated with the clathrin, and it has been suggested that each subpopulation of vesicles (each transporting specific contents to their correct location in the cell) *has its own group of unique accessory proteins*. These proteins have been termed the 'docking markers' of the vesicle and it is presumed that they are recognised by acceptor molecules on the cytoplasmic face of the target or destination membranes. It may be that coated vesicles are the principal means by which the addressing decisions of the intracellular traffic are accomplished.

References and further reading

Dautry-Varsat, A. and H. F. Lodish. 1984. How receptors bring proteins and particles into cells. *Scient. Am.* **250** (May), 48-54.

Edleman, G. M. 1984. Cell-adhesion molecules: a molecular basis for animal form. *Scient. Am.* **250** (Apr.), 80-91.

Lodish, H. F. and J. E. Rothman 1979. The assembly of cell membranes. *Scient. Am.* **240** (Jan.), 38-53.

Miller, K. R. 1979. The photosynthetic membrane. *Scient. Am.* **241** (Apr.), 100-11.

Satir, B. 1975. The final steps in secretion. *Scient. Am.* **233** (Oct.), 28-37.

Staehelin, L. A. and B. E. Hull 1978. Junctions between living cells. *Scient. Am.* **238** (May), 140-53.

Unwin, N. and R. Henderson 1984. The structure of proteins in biological membranes. *Scient. Am.* **250** (Feb.), 56-66.

Cell Biology

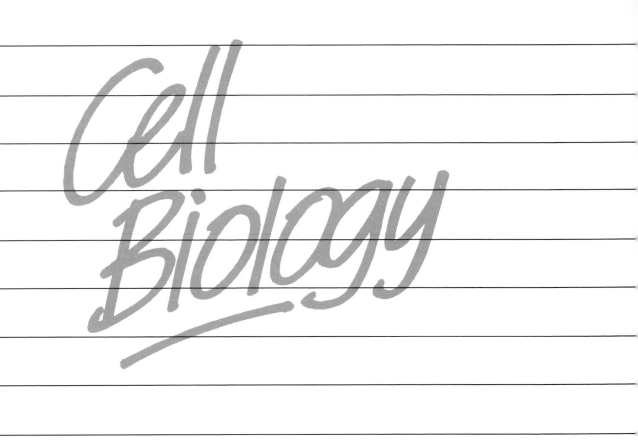

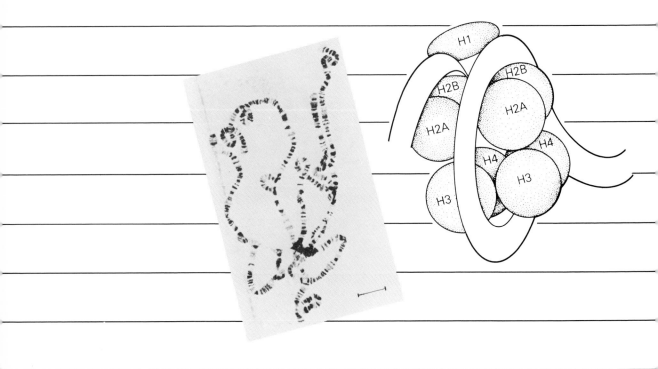

Molecular basis of heredity

M. R. Hartley

MODERN VIEWS IN BIOLOGY

5.1   Molecular biology

The aims of the branch of biology known as molecular biology are to explain broadly in molecular terms the two paramount characteristics of living organisms: *their ability to grow and their ability to reproduce themselves with a high degree of accuracy*. The discovery of the structure of DNA (deoxyribonucleic acid), the hereditary material, by Watson and Crick in 1953 provided a conceptual and mechanistic framework for molecular biology. In the past ten years or so, the pace of research and discovery has greatly increased, largely as a result of the application of **recombinant DNA techniques** (see §6.10). In fact, most of the major discoveries in molecular biology have only been made possible after the development of new techniques. As well as answering fundamental questions about how organisms grow and reproduce, exciting new applications of molecular biology are emerging, particularly in biotechnology.

5.1.1   GENES AND PROTEINS

Although the fact that many human traits, such as eye and hair colour, are passed on from parents to offspring has been known since the dawn of civilisation, the inheritance of such traits was not characterised until the progenies of controlled crosses between individuals with differing and well defined traits were analysed quantitatively. This was first accomplished in the 1860s by Gregor Mendel, whose work on the inheritance of morphological characteristics among the progenies of different, true breeding lines of peas established genetics as an experimental science. Mendel's work showed that units of heredity consisting of '**particulate factors**' are transmitted from parents to offspring. We now call these particulate factors **genes**. The physical basis of heredity was a complete mystery in Mendel's time, and it was not until the early part of this century that it was realised that the behaviour of **chromosomes** at meiosis and mitosis *accounted for the transmission and segregation of Mendel's particulate units of inheritance*. Formal proof that genes lie on chromosomes was provided in 1910 by Morgan, who showed that the inheritance of a mutant, white eye colour (a recessive mutation) in the fruit fly *Drosophila melanogaster* was dependent on the absence in male progeny of the X chromosome from the female which specified the normal (wild type) red eye colour.

   In the early part of this century, the biochemical basis of the morphological traits whose inheritance was studied and on which the science of genetics was founded, was completely unknown. Indeed, even today *most mutations which affect morphology remain to be characterised biochemically*, although significant advances have recently been made in this area. The main reason for the slow progress is that morphological mutants whose phenotypes can be attributed to a single Mendelian gene often show a variety of discernable differences when compared with the wild type, thus making it difficult to pinpoint the primary consequences of the mutant gene. The appreciation of this constraint in the 1940s led to the American geneticist Beadle and the biochemist Tatum to adopt a different approach. As Beadle recollected 'it suddenly occurred to me that it ought to be possible to reverse the procedure we had been following and instead of attempting to work on the chemistry of known genetic differences, we should be able to select mutants in which known chemical reactions were blocked'. The organism which Beadle and

Tatum chose to study was the bread mould *Neurospora crassa*. Wild type *N. crassa* has simple nutritional requirements and will grow on a simple, defined **minimal medium** containing inorganic salts, a carbohydrate source (e.g. glucose or sucrose) and the vitamin biotin (required for some carboxylation reactions). By inducing mutations with ultra-violet light, X-rays or chemical mutagens, mutants were selected which no longer grew on the minimal medium, *but required additional growth substances specific to each mutant type*. For example, some mutants required amino acids such as arginine or tryptophan while others required vitamins such as nicotinic acid or riboflavin. In each mutant, an enzyme involved in the synthesis of the required growth factor was shown to be absent (or at least inactive). An example will now be considered. Several different mutants of *N. crassa* which require arginine for growth have been isolated. These strains can be classified on the basis of the substitution that can be made for arginine in satisfying the growth requirement (Table 5.1). It can be seen that one strain (no. 36703) has a specific requirement for arginine, while other strains (30300 and 27947) grow equally well in the presence of either arginine or citrulline. Other strains (21502) grow when either arginine, citrulline or ornithine is added to the medium. The interpretation of these data is that *arginine is synthesised via citrulline and ornithine*. The enzymes catalysing the interconversions are under the control of independently mutatable genes, i.e.

ornithine ——gene 1—→ citrulline ——gene 2—→ arginine
ornithine transcarbamylase   arginosuccinic synthetase

From this type of experimentation it was postulated that the main function of genes is to control the synthesis of enzymes on a one for one basis, and gave rise to the **one gene one enzyme hypothesis**. Further evidence to support this idea came in 1957 when Ingram studied the molecular basis of **sickle-cell anaemia**, a lethal disease in homozygotes and common in people of African descent. The haemoglobin of adults is composed of two chains ech of α- and β-globins, folded around iron-containing haem groups (see §3.2.2). Ingram showed that the haemoglobin in sickle-cell anaemia patients (Hs) differed from that of healthy people in that the sixth amino acid residue of the β-globin chain, glutamic acid in normal haemoglobin, was replaced by valine in Hs.

Table 5.1 Growth of mutant strains of *Neurospora*. The values represent increases in dry weight in mg after five days in liquid medium supplemented with 0.005 mM arginine, ornithine or citrulline, or without supplement. (From Srb A. M. and N. H Horowitz 1944. *J. Biol Chem.* **154**, 133.)

Mutant strain no.	Arginine	Citrulline	Ornithine	Unsupplemented minimal medium
21502	37.2	37.6	29.2	0.9
27947	20.9	18.7	10.5	0.0
34105	33.2	30.0	25.5	1.1
30300	37.6	34.1	0.8	1.0
33442	35.0	42.7	2.5	2.3
36703	20.4	0.0	0.0	0.0

With increasing knowledge of enzyme structure, it became clear that many enzymes are composed of more than one type of polypeptide chain (as in the case of haemoglobin above) and that a mutation affecting one type of chain would not affect others but could abolish or modify enzymatic activity. Also, not all genes code for proteins (see §6.2.2). So the one gene, one enzyme hypothesis is not universally applicable and a more general statement is **one gene, one polypeptide chain**.

Implicit in the foregoing discussions is the fact that nearly all of the multitude of chemical reactions occurring within cells are catalysed by enzymes. Because enzymes exhibit a high degree of specificity towards their substrates (see §3.3), it follows that the number of different enzymes within a cell is very great indeed. Even a relatively 'simple' prokaryotic cell contains in the region of 2000-3000 different enzymes and vertebrates probably contain 10 000-20 000, though not all of these would be present in every cell or tissue type because of specialisation. The catalytic activity of an enzyme is determined by the precise three-dimensional folding of its polypeptide chain(s) (see §3.4). The folding is dependent on the weak interactions between the R groups of the constituent amino acid residues. The finding that denatured (inactive) enzymes can, under appropriate conditions, renature and restore full enzymatic activity means that ultimately only the nature and order of the amino acid residues of the polypeptide chain (i.e. the primary structure) are the determinants of an enzyme's activity. *Thus the main function of the genetic material is to specify the primary structure of a large number of different polypeptides.*

5.1.2   GENETIC MATERIAL

Although eukaryotes had been used to establish that genes lie on chromosomes, and provide the evidence for the one gene, one polypeptide hypothesis it was through studies on prokaryotes – bacteria and their viruses – that the identity of the genetic material was established. In 1928 the British microbiologist Griffith studied the basis of pathogenicity of the bacterium *Diplococcus pneumoniae* (also known as *pneumococcus*) which causes pneumonia. He observed that the cells of pathogenic strains (termed S because they form smooth colonies) were surrounded by a thick slimy polysaccharide outer coat – the capsule – whereas mutant strains which lacked the capsule were non-pathogenic (termed R because they form rough colonies). Griffith made the very surprising observation that *if the non-pathogenic R strain bacteria were mixed with heat-killed S cells and injected into mice, the mice became infected and died.* Neither heat-killed S strains nor viable R strains alone resulted in pathogenic symptoms. Furthermore, viable, encapsulated S cells were isolated from the blood of the dead mice. Thus, the heat-killed S cells had somehow **transformed** live R cells into live S cells, i.e. *a heritable genetic change had occurred.* The factor responsible for transformation, termed the **transforming principle**, was not identified by Griffith. This task was taken up in 1944 by Avery and his colleagues who prepared cell-free extracts of the S bacteria, fractionated the extracts into various components and assayed these for their ability to transform R cells into S cells *in vitro* (Fig. 5.1). They found that *the transforming principle resided in DNA.* Some of the evidence used to confirm this was: (a) incubation of the extract with deoxyribonuclease, an enzyme that specifically degrades DNA, destroyed any transforming activity; and (b) the proteolytic enzymes

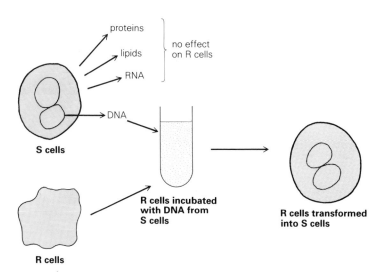

Figure 5.1 Evidence that the transforming principle is DNA. Of the various components isolated from *Diplococcus* S cells, only DNA has the ability to transform R cells into S cells.

trypsin and chymotrypsin (which degrade all proteins) did not affect this transforming activity.

Following the publication of Avery's findings, many scientists were sceptical of the conclusion that DNA was the genetic material, largely because the structure of DNA was not known at that time and it was widely believed that DNA was a relatively simple, small molecule without the necessary complexity and specificity expected of the genetic material. However, further support for the genetic role of DNA was provided by studies on a virus (bacteriophage) which infects the colon bacillus *Escherichia coli*. Phage T_2 consists of a core of DNA surrounded by a protein coat which is differentiated into head, sheath and contractile tail (Fig. 5.2). Upon collision with a susceptible host cell, the phage becomes attached to the cell wall via its tail and enzymatically creates a small hole in the cell wall through which it injects its **genome** (genetic material) into the cell. The phage genome then replicates within the cell, making 100-200 new copies which become encapsulated within newly synthesised protein coats. The bacterial cell wall then breaks open (**lyses**), releasing the progeny phage into the surrounding medium. The question as to whether it is the DNA, or one of the protein components, which specifies the information for the synthesis of new phage particles was answered in 1952 by the American geneticists Hershey and Chase. They prepared phage labelled with both radioactive phosphorus (^{32}P), a constituent of DNA but not protein, and radioactive sulphur (^{35}S), a constituent of protein (in cysteine and methionine residues) but not DNA. The labelled phage particles were made by infecting *E. coli* cells in a medium containing ^{32}P-orthophosphate and ^{35}S-sulphate and recovering the progeny phage. They then performed the experiment shown in Figure 5.3. The result was that *most of the labelled DNA entered the bacterial cell where it directed the synthesis of progeny phage, whereas most of the labelled protein remained on the outside of the cell and was recovered in the supernatant.* Hence the genetic material of phage T_2 was shown to be DNA.

In recent years a large and convincing body of evidence has accumulated to show that DNA is also the genetic material of eukaryotes. However, in certain viruses which infect animals, plants and bacteria, DNA is replaced by the chemically similar ribonucleic acid (RNA) as genetic material.

Figure 5.2 Bacteriophage T_2. The icosohedral head with 30 facets encapsulates the chromosome (DNA). It is attached by a neck to a contractile sheath composed of protein sub-units surrounding a hollow cone, and ends in a spiked end plate to which six fibres are attached. During infection of a susceptible host bacterium, the spikes and fibres attach to the host's cell wall, the sheath contracts, driving the core through the wall, and the phage DNA enters the cell.

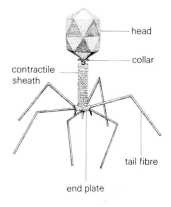

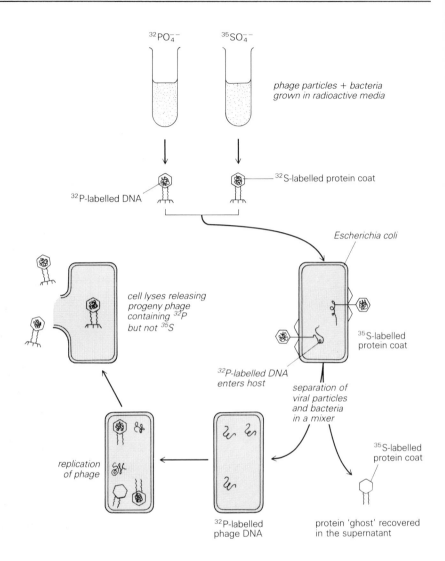

Figure 5.3 The Hershey and Chase experiment (see text).

5.2 The structure of nucleic acids

There are two types of nucleic acids found within cells and viruses: DNA and RNA. DNA is a major component of chromosomes and as such is largely confined to the nucleus. It is also found in smaller quantities in the mitochondria of animal cells and the mitochondria and plastids (chloroplasts, etc.) of plant cells (see §2.6.3). RNA is found in the nucleus, mitochondria, plastids and cytoplasm. Both DNA and RNA are large, polymeric molecules consisting of a long series of **nucleotides**. Each nucleotide comprises a **phosphate group** linked to a **pentose sugar** (ribose in RNA, deoxyribose in DNA) which in turn is linked to either a **purine** or **pyrimidine** base (flat ring-shaped molecules containing nitrogen) (Fig. 5.4a). The purine bases within both DNA and RNA are **adenine** (A) and **guanine** (G): the pyrimidine bases in DNA are **thymine** (T) and **cytosine** (C). In RNA, thymine is replaced by the structurally similar **uracil** (U) (Fig. 5.4b). In **polynucleotides** (i.e. nucleic acids) the nucleotides are linked together by a backbone of pentose residues alternating with phosphate groups *such that the 5' position of one pentose ring is connected to the 3' position of the adjacent pentose by a phosphate group*. This forms the 5'-3' **phosphodiester linkage** (Fig. 5.4c). The polynucleotides also possess **polarity** in that one terminal pentose residue bears a 5' phosphate group (referred to as the 5' end) while the other terminal pentose bears a free 3' hydroxyl group (referred to as the 3' end).

Shorthand notation of a short nucleic acid is shown below:

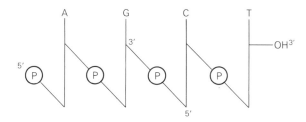

or ^{5}pApGpCpT—OH^3 or simply ^{5}AGCT^3. It is conventional *to write nucleic acid sequences with the 5' terminus on the left and the 3' terminus on the right*.

5.2.1   DNA IS A DOUBLE HELIX

DNA is the single most important molecule found within cells and the analysis of its structure in 1953 by Watson and Crick was arguably the most significant biological finding of this century. Watson and Crick combined the available chemical and physical data for DNA with the X-ray diffraction patterns obtained by Wilkins (see §1.9). The X-ray diffraction in analysis was performed on fibres of the so-called B form of DNA where the individual strands pack together regularly. The essential features of the Watson-Crick DNA model are shown in Figure 5.5 and summarised below.

(a) The two helical polynucleotide chains are coiled around a common axis. The two chains have opposite polarity, i.e. they are **antiparallel**, one running from the 5' to the 3' end and the other from the 3' to the 5' end.

(b) The regular repeating sugar-phosphate backbone of each strand lies on the outside of the helix. The purine and pyrimidine bases project inwards at 90° to the axis of the helix.

(c) The two strands are held together by hydrogen bonding between pairs of bases such that guanine always pairs with cytosine and adenine always pairs with thymine; this is called **complementary base pairing**. This feature of the structure is in accordance with the findings of Chargaff that irrespective of the actual base composition of the DNA (which can vary widely between organisms) the *proportions of A and T are always equivalent, as are those of C and G* (i.e. [A] = [T] and [G] = [C], where [ ] denotes molar concentration). It thus follows that [purines] = [pyrimidines]. Because each base pair contains one two-ringed purine

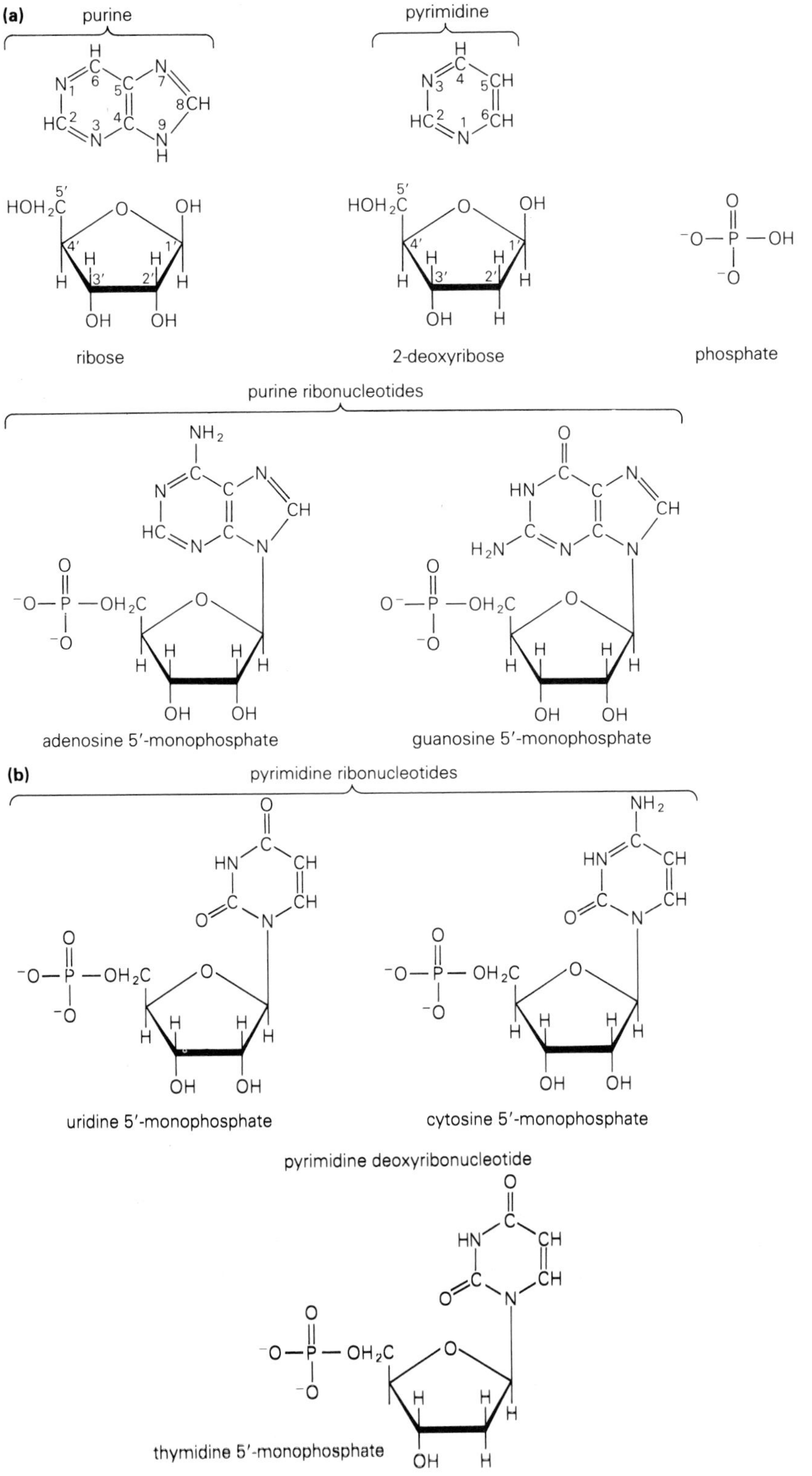

(a) purine pyrimidine

ribose 2-deoxyribose phosphate

purine ribonucleotides

adenosine 5'-monophosphate guanosine 5'-monophosphate

(b) pyrimidine ribonucleotides

uridine 5'-monophosphate cytosine 5'-monophosphate

pyrimidine deoxyribonucleotide

thymidine 5'-monophosphate

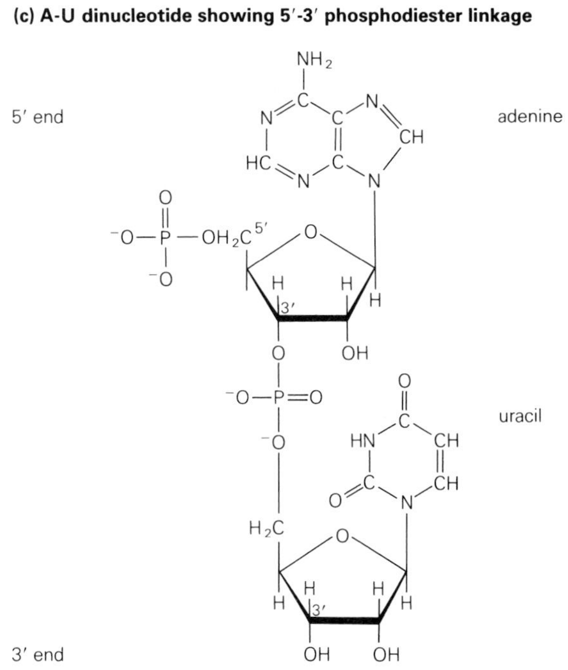

(c) A-U dinucleotide showing 5′-3′ phosphodiester linkage

5′ end   adenine

3′ end

Figure 5.4 The chemical constituents of nucleic acids. The corresponding deoxyri-
bonucleotides lack the oxygen atom at carbon atom 2 on the pentose sugar.

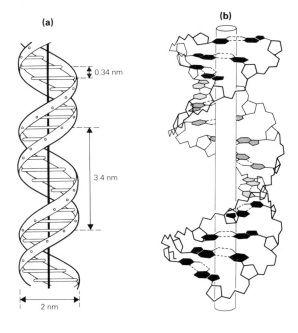

(a)

0.34 nm

3.4 nm

2 nm

(b)

Figure 5.5 The structure of
DNA. (a) Diagrammatic
representation of the DNA
molecule as proposed by Watson
and Crick. The two
sugar–phosphate chains are
shown as ribbons and the base
pairs as horizontal rods.
(b) Skeletal model of DNA. The
dotted lines represent hydrogen
bonds between the base paris
(shaded).

and one single-ringed pyrimidine the dimensions of all the base pairs are roughly the same so that *the helix has a constant diameter*. If purine:purine or pyrimidine:pyrimidine partnerships occurred the helix would become distorted. The structure of the hydrogen-bound base pairs is shown in Figures 5.6 and 5.7. Note that *three hydrogen bonds* can form between G:C base pairs (symbolised by G≡C) but only *two can* form between an A:T base pair (symbolised by A=T).

(d) The diameter of the helix is 2.0 nm and adjacent bases are separated by 0.34 nm and inclined at 36° relative to each other. This means that each complete turn of the double helix contains about ten base pairs.

(e) The actual **sequence** of the bases along a particular stretch of DNA is in no way restricted by stereochemical considerations and it is this precise sequence which carries genetic information. As a consequence of the base-pairing rules it follows that one strand of the helix must be complementary in sequence to the other strand. For example, if one strand has the sequence ^{5}TCGTCA^3 the other strand necessarily has the sequence ^{3}AGCAGT^5.

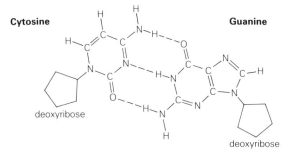

Figure 5.6 Hydrogen bonding between adenine–thymine and guanine–cytosine base pairs.

The Watson-Crick model of DNA structure has proved to be essentially correct and is consistent with a vast amount of accumulated information. However, it should be pointed out that the parameters used by Watson and Crick in their model-building exercises were derived from X-ray diffraction studies on fibres in which it is not possible to assign positions to specific atoms. As a consequence, *the model reflects an average structure, on which subtle variations may exist in vivo for particular regions*. An alternative model which has received considerable attention in recent years is the so-

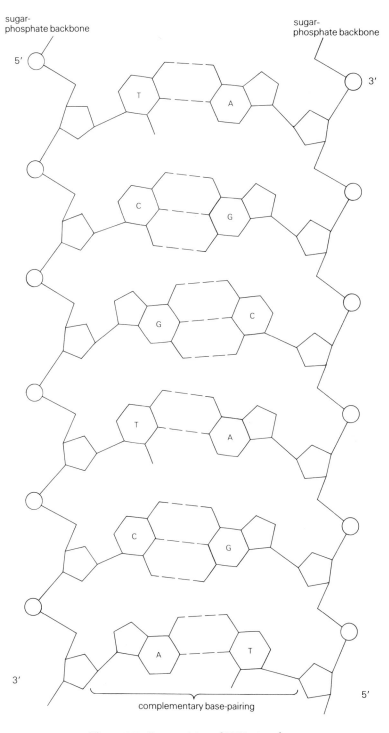

sugar-
phosphate backbone

5'

sugar-
phosphate backbone

3'

T — A

C G

G C

T — A

C G

A --- T

3'

5'

complementary base-pairing

Figure 5.7 Base–pairing of DNA strands.

called **Z-DNA model** in which the helix is left-handed, contains 12 base pairs per turn and derives its name from the zig-zag path followed by the sugar-phosphate backbone along the helix.

As we shall see, DNA molecules can be very long indeed, but it is important to realise that *in vivo* they do not exist in an extended linear form. For example, *human chromosomes probably contain a single DNA molecule which in its extended form would be about 1-7 cm in length. However, condensed metaphase chromosomes have lengths of 2-10 μm indicating that the degree of condensation within the chromosome is in the order of 7000-fold!* In part, this condensation results from the association of DNA with a set of basic proteins, the **histones** (see §5.4).

5.3   Replication of DNA

As was stated in the introduction, one characteristic of living organisms is that they are capable of reproducing themselves with a high degree of accuracy. It follows that the genetic material, DNA, must also possess this quality. In their classic paper on DNA structure, Watson and Crick stated that 'it has not escaped our notice that the specific (base) pairing we have postulated immediately suggests a possible copying mechanism for the genetic material'. The idea was that the two complementary strands could separate from each other by disruption of the relatively weak forces joining them (hydrogen bonds) and because of the specificity of base pairing each strand could then function as a **template** for the synthesis of a new complementary strand. Experimental support for this notion was provided by Meselson and Stahl in 1958 using the newly developed technique known as **density labelling**. *E. coli* cells were grown for many generations in a medium containing the heavy stable isotope of nitrogen, ^{15}N (in the form of $^{15}NH_4Cl$), such that all the nitrogen atoms of the bases contained ^{15}N, thereby making the DNA more dense than ordinary DNA. If such 'heavy' DNA is mixed with normal 'light' DNA and centrifuged to equilibrium in a concentrated solution of caesium chloride, the DNA molecules redistribute in the gradient of caesium chloride concentration which forms, and band at positions in the gradient at which the density of the solution is equal to their own buoyant density (see §1.5.2). Hence two DNA bands of different buoyant density are resolved, corresponding to the 'heavy' and 'light' DNAs (Fig. 5.8). The ^{15}N-labelled cells were transferred to a normal medium containing ^{14}N and allowed to continue to grow. Samples of the bacteria were taken at different times during the experiment, DNA extracted from them and their buoyant densities determined in the ultracentrifuge. The results are shown in Figure 5.8. As can be seen, the cells grown entirely in ^{15}N-containing medium contain all 'heavy' DNA. After one cell generation following transfer to ^{14}N medium the DNA has a density intermediate between that of 'heavy' DNA and 'light' DNA. After two generations, two bands are visible, a 'light' band and a band of intermediate density. In successive generations the proportion of 'light' DNA increases relative to that of intermediate density DNA. The interpretation of these results is that the replication of DNA is **semi-conservative**, that is that the two parental DNA strands serve as templates for the synthesis of complementary daughter strands as predicted by Watson and Crick (Fig. 5.9).

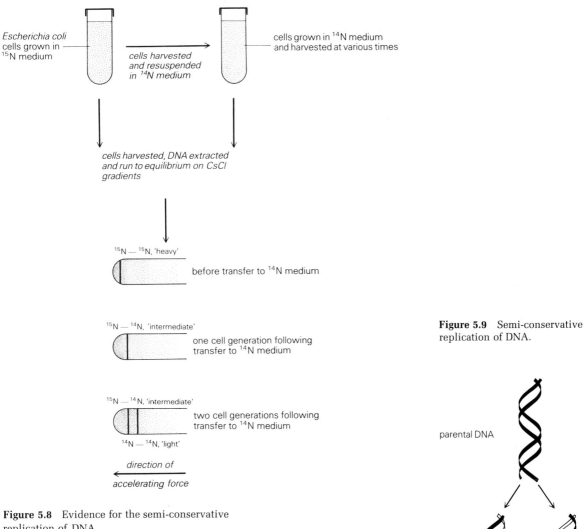

Figure 5.8 Evidence for the semi-conservative replication of DNA.

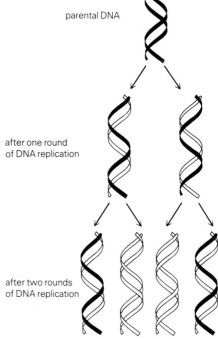

Figure 5.9 Semi-conservative replication of DNA.

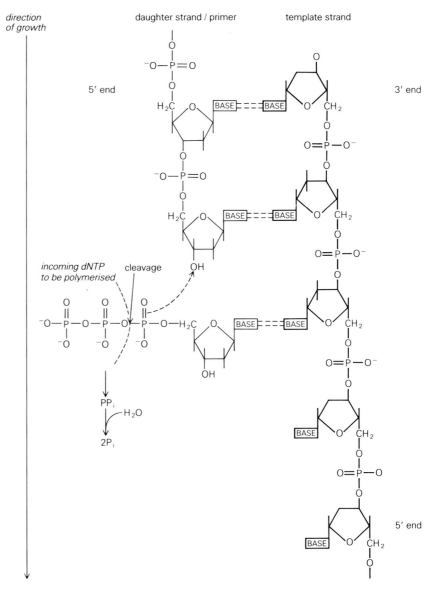

Figure 5.10 The synthesis of DNA by DNA polymerase. The incoming dNTP is aligned by specific base-pairing with the complementary base in the template strand before formation of the 5′–3′ phosphodiester linkage ensues.

5.3.1 MECHANISM OF REPLICATION

In 1956 Arthur Kornberg discovered that cell-free extracts of E. coli contained an enzyme, DNA polymerase I, capable of synthesising DNA from its precursors. The main features of the activity of this enzyme are:

(a) The DNA precursors must be in the form of *deoxyribonucleotide 5′-triphosphates* and all four of these (dATP, dCTP, dTTP and dGTP, known collectively as dNTPs) must be present.

(b) The enzyme adds dNTPs to the 3'—OH terminus of a pre-existing DNA (or RNA) strand, known as the **primer**. During this reaction a new phosphodiester linkage is formed between the 3'—OH terminus of the primer and the innermost phosphate group of the dNTP, liberating pyrophosphate. The subsequent hydrolysis of pyrophosphate provides the energy to drive the reaction.

(c) The polymerase takes its instructions from a single-stranded **DNA template** which is essential for activity. The dNTPs align themselves on the DNA template by complementary base pairing which follows the Watson-Crick rules. Thus *the precise sequence of the newly synthesised strand is determined by the template strand* (Fig. 5.10).

Following the discovery of DNA polymerase I, it was assumed that this enzyme was chiefly responsible for replicating DNA. However, mutants of *E. coli* which were almost totally lacking DNA polymerase I activity (termed **pol A1** mutants) were isolated, *and shown to replicate their DNA at a rate comparable to the wild type*. It is now known that another enzyme, DNA polymerase III, is mainly responsible for polymerisation. The pol A1 mutants

Figure 5.11 (a) Replication of the circular *Escherichia coli* chromosome. Replication starts from a fixed origin and proceeds bidirectionally, resulting in two replication Y forks (indicated by the small arrows). Parental strands are shown as continuous lines, daughter strands as dotted lines. (b) EM autoradiograph of the chromosome of *E. coli* in the process of replication. The DNA is visible because it has incorporated tritiated thymine for two generations. An interpretation is shown in the inset. The double solid lines represent duplex DNA in which both strands are labelled. The single solid single dashed lines represent duplex DNA in which only one strand is labelled. X and Y show the positions of the replicating Y-forks. The bar represents 100 μm.

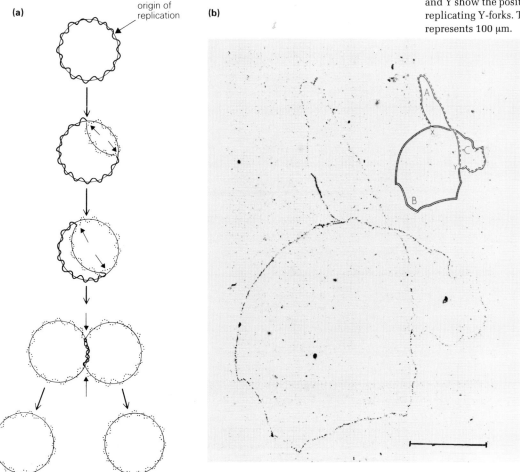

(a) origin of replication **(b)**

were shown to be far less capable of repairing damage to DNA sustained by ultra-violet light (see §5.3.2) than the wild type. *The main functions ascribed to DNA polymerase I are DNA repair (see §5.3.2) and the filling in of gaps between the precursor fragments of DNA (see below).*

Autoradiography (see §1.8.4 and Figs 5.11a & b) of replicating prokaryotic chromosomes labelled with ^{3}H-thymidine, a precursor of dTTP, has shown that DNA replication is confined to regions called **replicating Y-forks**, so-called because of their Y-shaped structure. As DNA replication progresses, the Y-forks move along the parental DNA helix. *If DNA synthesis were to occur continuously on both of the single-stranded template strands made available for the polymerase at the Y-fork, one daughter strand would grow in a 5' to 3' direction while the other would have to grow in a 3' to 5' direction (because the parental strands are antiparallel). However, DNA polymerase can only synthesise new strands in the 5' to 3' direction.* The solution to this problem is achieved in the following way:

(a) The so-called leading strand is synthesised continuously in a 5' to 3' direction.

(b) The so-called lagging strand is synthesised in short stretches (1000-2000 nucleotides long in bacteria) again in a 5' to 3' direction, *thereby creating a discontinuous molecule containing gaps. These gaps are later 'filled in' by DNA polymerase I and the continuity of the molecule is restored by sealing the ends between adjacent stretches of DNA with an enzyme called DNA ligase.* In fact, DNA polymerase III cannot initiate DNA chains *de novo* but requires a primer molecule. This primer is RNA synthesised by an RNA polymerase called **primase**. DNA polymerase III then extends this RNA primer as a DNA strand. The RNA primer is subsequently degraded and the gap filled in by DNA polymerase I (Fig. 5.12).

It is axiomatic for DNA synthesis that single-stranded DNA is provided at the replicating Y-fork. However, the complementary DNA strands are held together quite strongly by hydrogen bonds, as shown by the fact that double-stranded DNA must be heated in water to about 90 °C before strand separation occurs. *In vivo* strand separation is accomplished by an enzyme called helicase which utilises the energy released by ATP hydrolysis. The *single-*

Table 5.2   Some of the enzymes involved in DNA replication in *Escherichia coli.*

Enzyme	Function
DNA polymerase III	the major enzyme responsible for DNA replication
DNA polymerase I	fills in gaps between the fragments of DNA synthesised by DNA polymerase III
helicase	uses energy released by ATP hydrolysis to disrupt hydrogen bonds between base pairs and thereby creates single-stranded regions
DNA binding protein	stabilises single-stranded DNA, preventing re-formation of the double helix
topoisomerase	unwinds the double helix
DNA ligase	uses ATP to seal breaks in sugar-phosphate backbone
RNA primase	produces short RNA sequences to prime DNA polymerase III

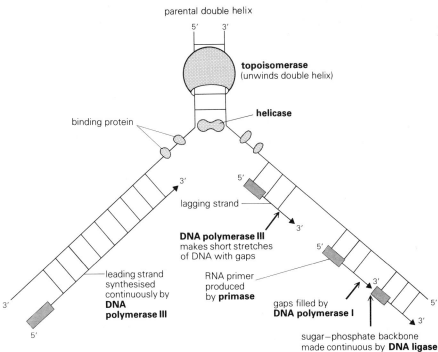

Figure 5.12 Some of the enzymatic events at the replicating Y-fork of *Escherichia coli*.

stranded DNA is then prevented from re-forming a double helix because it is stabilised by specific binding proteins. A topological problem the cell has to overcome arises in the unwinding of the double helix around the replicating Y-fork (because the DNA strands are helically intertwined, they must unwind to replicate, a problem compounded by the fact that many DNAs are in the form of covalently closed circles). To account for known rates of DNA replication it has been estimated that the DNA strands must unwind by revolving around their long axis at *100 revolutions per second.* Flailing of the 'arms' of the replicationg Y-fork is reduced by the continual breaking and re-forming of phosphodiester bonds, a process catalysed by a class of enzymes known as **DNA topoisomerases**. Some of the important enzymes involved in DNA replication are summarised in Table 5.2.

5.3.2   ACCURACY OF DNA REPLICATION

DNA replication is an exceedingly accurate process, and it has been estimated that *only one error is made in about 10^9 base pair replications.* This high fidelity is due, in large part, to the finding that DNA polymerase has a *proofreading activity* in addition to its polymerising activity. One such proofreading activity results from the fact that DNA polymerase can hydrolyse phosphodiester bonds in single strands if mismatched regions occur at the 3'—OH terminus of the primer strand. *Thus an incorrect base is removed before further polymerisation occurs* (Fig. 5.13).

DNA is continually sustaining damage from a variety of physical and chemical agents such as ionising radiation, ultra-violet light and chemical carcinogens (agents that cause cancers). Damage includes the introduction of

Figure 5.13 Proofreading activity of DNA polymerase.

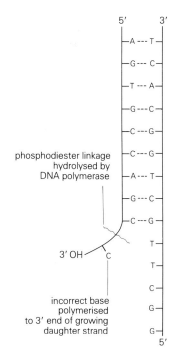

(a)

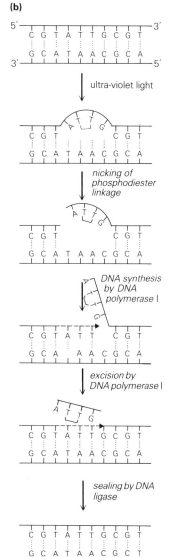

Figure 5.14   (a) Formation of a thymine dimer by the action of ultra-violet light. Adjacent thymine residues are covalently joined by the bonds shown. The C=C bond of each thymine ring is converted to a C–C bond in each cyclobutyl ring.
(b) Excision and repair of a region of DNA containing a thymine dimer.

(b)

single-stranded breaks, loss or alteration of bases and covalent cross-linkage between bases in the same strand. Very few of these changes accumulate because all cells are able to correct the lesions by a process known as **DNA repair**. Consideration of a well documented example will make this clear. Adjacent pyrimidine residues on the same strand can become covalently linked on exposure to ultra-violet light to form a **pyrimidine dimer** (Fig. 5.14a). Such a dimer *disturbs the double helix and blocks replication and gene expression.* Pyrimidine dimers are corrected by an **excision-repair** system which removes the dimerised bases and then synthesises a new stretch of DNA to replace them (Fig. 5.14b). Although details of the excision-repair system were worked out using *E. coli*, a similar system operates in humans. The rare skin disease xeroderma pigmentosum, in which homozygous individuals are extremely sensitive to ultra-violet light and develop skin cancer, is caused by a defect in the enzyme which hydrolyses the DNA backbone near a pyrimidine dimer.

5.3.3   MUTATION

Mutant organisms are defined in genetic studies as showing altered phenotypes relative to the wild type. In biochemical terms, *the altered phenotype is the result of a change in the amino acid sequence of a protein(s).* If the altered amino acid(s) occurs at a position in the protein molecule at or near the active site or causes a conformational change in the protein then the enzymatic activity of that protein will be impaired or altered. An example of this, the mutant form of haemoglobin HbS which causes sickle-cell anaemia, has already been discussed (§5.1.1, see also §3.2.2). Not all mutations have deleterious effects. Many mutations are described as *neutral* in that they do not affect the '*fitness*' of an organism in any obvious way. However, the accumulation of neutral mutations is probably very important in evolutionary terms, since such mutations provide the variation upon which natural selection can act.

Mutations arise through changes in the base sequence of particular genes. In theory, any base pair in DNA is capable of being changed. Some mutations, called **spontaneous mutations**, occur naturally at a characteristic rate (usually about 1×10^9 for any one base pair per cell generation) for a particular organism and arise both through rare mistakes in DNA replication and damage sustained by a variety of intracellular and environmental chemical and physical agents. The frequency of mutation can be greatly increased by exposure of cells or organisms to certain compounds called **mutagens**, and such changes are called **induced mutations**. Examples of chemical mutagens include base analogues such as 5-bromouracil, which is incorporated into DNA and alters base pairing during subsequent replication, nitrous acid, which oxidatively deaminates the amino groups of adenine, cytosine and guanine, again altering base pairing, and **intercalating** agents such as the acridines, which slip between adjacent bases in DNA and result in the insertion or deletion of one or more base pairs. Many carcinogens are also mutagenic, and a great deal of effort is going into identifying mutagens in the environment so that human exposure to them can be minimised.

Several different types of mutations are known: alterations which change single base pairs are termed **point mutations** which can be of two forms: (a) **transitions**, in which one purine is substituted by another purine (A replaces G or vice versa) or one pyrimidine is substituted by another pyrimidine (C replaces T or vice versa), or (b) **transversions** in which a purine is replaced by a pyrimidine or vice versa. Other types of mutation include the **insertion** or **deletion** of one or more base pairs. An example of a transition will be considered. One effect of nitrous acid is to deaminate cytosine to form uracil (Fig. 5.15). When the next round of DNA replication ensues, uracil base pairs with adenine (instead of with guanine with which the original cytosine was base paired).

5.4   Complexity and organisation of DNA

The total amount of DNA per haploid genome of an organism is known as its **C value** and is characteristic of that species. (There are a few interesting exceptions, for example the flax plant, in which the C value can vary by 10 per cent depending on nutritional conditions.) The C value may be expressed as pg DNA (pg = 10^{-12} g) or base pairs (bp) (1 pg DNA is equivalent to 0.965×10^9 bp). The C values of groups of organisms of widely differing phylogenies are shown in Table 5.3. As can be seen, they vary greatly – from about 10^6 bp for *Mycoplasma* (a genus of Gram negative parasitic prokaryotes, see §2.5.1) to about 10^{11} bp for some amphibians and higher plants. Although it is evident that there is a general increase in C value with increasing evolutionary complexity, this relationship breaks down within some classes. For example, amphibians show a 100-fold variation in their C values and even different species of the same genus can show a 10-fold variation. It is very unlikely that such *large differences in C values between closely related organisms reflect corresponding differences in the numbers of different genes they possess*. It is true to say that higher organisms contain far more DNA than can be accounted for in terms of that required to code for proteins. These imponderables are embodied in the so-called **C value paradox** which refers to our inability to explain genome sizes in terms of expected functions!

Within prokaryotic genomes each gene is usually represented only once per genome. The total genome is contained within a circular chromosome. In eukaryotes, genome organisation is far more complex. A sizable proportion of the genome (10-80 per cent) consists of families of **repetitive sequences**, the individual members of which are related but not identical in sequence. Repetitive sequences may be either **tandemly arranged** in blocks, or may be interspersed with **single copy sequences** represented once or a few times per

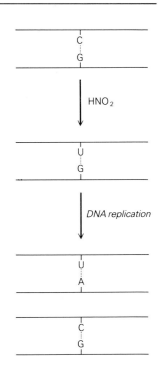

Figure 5.15 Mutation induced by nitrous acid.

Table 5.3 C values (haploid DNA content in base pairs) of various groups of organisms. (From *Genes*, B. Lewin, copyright © 1983 John Wiley & Sons, Inc. Reprinted by permission of John Wiley & Sons, Inc.)

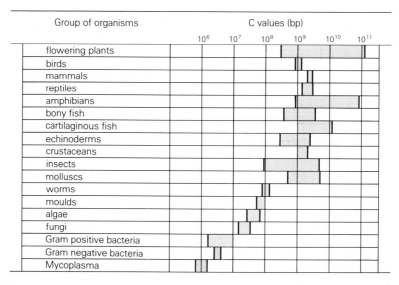

Group of organisms	C values (bp)
	10^6 10^7 10^8 10^9 10^{10} 10^{11}
flowering plants	
birds	
mammals	
reptiles	
amphibians	
bony fish	
cartilaginous fish	
echinoderms	
crustaceans	
insects	
molluscs	
worms	
moulds	
algae	
fungi	
Gram positive bacteria	
Gram negative bacteria	
Mycoplasma	

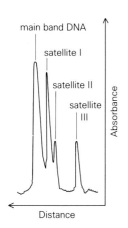

main band DNA

satellite I

satellite II

satellite III

Absorbance

Distance

Figure 5.16 Densitometer scan of a caesium chloride gradient fractionation of *Drosophilia virilis* DNA. Note the three prominent satellite bands.

Figure 5.17 The β-globin gene cluster in rabbit. The β₃ and β₄ genes, whose sequences differ slightly from that of the β₁ (adult) gene, are expressed during embryonic development, but not in the adult.

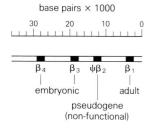

base pairs × 1000

30   20   10   0

β₄   β₃   ψβ₂   β₁
embryonic   adult
pseudogene
(non-functional)

haploid genome. Examples of clustered (tandemly arranged) repetitive sequences are those of **satellite** DNA bands found in many eukaryotes and seen as minor components in caesium chloride density gradient centrifugation. For example *Drosophila virilis* DNA has three satellite bands of lower buoyant density than the major component (Fig. 5.16). Each satellite band consists of several million copies of a repeating heptanucleotide sequence:

satellite I   satellite II   satellite III
5'ACAAACT3'   5'ATAAACT3'   5'ACAAATT3'
3'TGTTTGA5'   3'TATTTGA5'   3'TGTTTAA5'

It is generally believed that most, but not all, of the genes coding for proteins reside in the single copy DNA. However, recent evidence suggests that many genes which we can regard as residing in single copy sequences are descended by duplication and variation from an ancestral gene, and the related genes are called a **gene family**. For example, the human genes coding for the α- and β-globin chains of haemoglobin are organised into clusters and include several functional genes as well as non-functional **pseudogenes** (Fig. 5.17, see also §3.2.2).

A given family of repeated sequences may have a single chromosomal location, or may be found on several or all of the chromosomes. While the function of some repeated sequences is known, for example the genes coding for ribosomal RNA and histone proteins (see §5.4.1 & §6.2.1), most are of unknown function. It has been suggested that some of the interspersed repetitive sequences may function in regulating the activities of protein coding genes which lie near them. Some of the tandemly arranged repetitive sequences are confined to centromeric regions (see *Reproductive biology* Chapter 1) and *may function in aligning chromosomes during meiosis and mitosis.*

Recent research involving DNA sequencing has shown that the genomes of both prokaryotes and eukaryotes are far less static in terms of their DNA sequences than was envisaged a few years ago. One mechanism which allows movement of sequences within the genome involves **transposable elements** or **transposons**. These elements specify the information necessary to transpose a duplicated copy of themselves from their original 'donor' site to a new target site elsewhere in the genome. Such transpositions can have profound effects on the development of the organism concerned. For example, in some strains of *Drosophila* a transposable element termed 'P' generates copies which become inserted into the chromosomes of developing embryos, causing the latter to fail to produce viable germ line cells, a phenomenon known as **hybrid dysgenesis**. Many other examples are known and *it is now believed that mobile genetic elements are common features of most organisms.*

5.4.1 STRUCTURE OF CHROMOSOMES

It has already briefly been mentioned that *in vivo* DNA is associated with proteins and is highly compacted so that chromosomes can fit within the confines of the nucleus (in eukaryotes) or the nucleoid in prokaryotes. The major distinguishing feature between nucleus and nucleoid is that the former contains several chromosomes within a membrane-bound organelle whereas the latter contains a single chromosome and is not membrane-bound. The diploid human nucleus contains 46 chromosomes with a *total amount of DNA equivalent to a length of 1.8 m, yet this is accommodated within a nuclear envelope of only 6 μm diameter.* To begin to understand how DNA packaging is achieved, we must first look at chromosome structure.

Eukaryotic chromosomes are composed of chromatin, which consists chemically of about 60 per cent protein, about 35 per cent DNA and perhaps 5 per cent RNA. The DNA of each chromosome consists of a single, very large, linear molecule. The protein component can be broadly divided into two classes: histone and non-histone. The latter class represents a multitude of different proteins including DNA polymerase and RNA polymerase. The histones are very well characterised, and are among the most abundant proteins of the cell. They are relatively small in size and are very rich in the basic amino acids arginine and lysine, which together account for about a quarter of the total amino acid residues (Table 5.4). The histones are therefore positively charged at neutral pH and combine with negatively charged DNA (due to the negative charge in phosphate groups) to form DNA-histone complexes. The histones designated H2A, H2B, H3 and H4 are components of nucleosomes and have highly conserved primary structures: for example, histone H4 of peas and cows differs at only two amino acid residues.

Electron microscopy of extended chromatin fibres reveals a 'beads on a string' appearance, the 'beads' being the nucleosomes and the 'string' linker DNA (Fig. 5.18). Brief incubations of chromatin with certain nucleases (enzymes which degrade nucleic acid) result in preferential degradation of the linker DNA, liberating the nucleosomes as monomeric units. Analysis of such nucleosomes has revealed the following: (a) the basic repeat length of the DNA associated with each nucleosome is about 200 bp; (b) complete digestion of the linker DNA reveals that 140 bp of DNA is associated with the nucleosome core, in which the DNA is wound around the outside of an octameric protein disc consisting of two copies of each of histones H2A, H2B, H3 and H4 – the dimensions of this core are 11 $\times$ 5.5 nm; and (c) histone H1

Figure 5.18 EM of a disrupted rat thymus nucleus. A few chromatin fibres, showing nucleosomes, are stretched out from a region of higher concentration. The EM was prepared using negative staining and the bar is 0.1 μm.

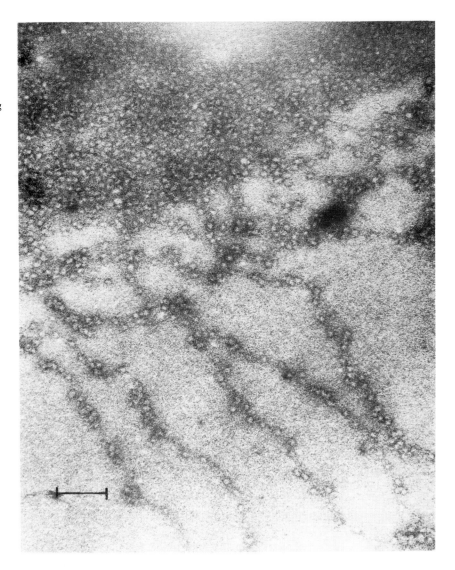

Table 5.4 Some properties of bovine histones. (From *Genes*, B. Lewin, copyright © 1983 John Wiley & Sons Inc. Reprinted by permission of John Wiley & Sons, Inc.)

Histone type	Basic amino acids (%)		Acidic amino acids (%)	Ratio basic : acidic	Molecular weight
	Lysine	*Arginine*			
H1	29	1	5	5.4	23 000
H2A	11	9	15	1.4	13 960
H2B	16	6	13	1.7	13 774
H3	10	13	13	1.8	15 342
H4	11	14	10	1.5	11 282

is associated with linker DNA, and may serve to pack nucleosomes together (Fig. 5.19).

In vivo chromatin rarely adopts the extended 'beads on a string' form, but instead is present in a more highly condensed form, commonly known as the **30 nm chromatin fibre**. The precise way in which the nucleosomes are ordered in 30 nm fibres is unknown, though several models have been proposed. One of these is shown in Figure 5.20c. This model would provide for a **DNA packing ratio** (degree of condensation relative to extended DNA) of about 40. *The actual packing ratio in human metaphase chromosomes is about 10^4*, so that 30 nm fibre itself must be further contorted into a higher order of structure. A schematic illustration of the various orders of chromatin organisation is shown in Figures 5.20a-f.

Through studies involving DNA sequencing, we now have fairly detailed knowledge about the organisation of genes in terms of the primary structure of their DNA. However, we know far less about how genes are organised within chromosomes. The most detailed information available has been provided by studies on two specialised types of chromosomes: **polytene chromosomes** and **lampbrush chromosomes**. Polytene chromosomes are found in the cells of certain tissues of Dipteran insects, most notably in the salivary glands of the larvae, and have been most extensively studied in *Drosophila* and *Chironomous* (midge). They are greatly enlarged relative to 'normal' chromosomes and arise through repeated cycles of chromatid replication without segregation of daughter chromatids since neither nuclear division nor cell division occurs. For example, in *D. melanogaster* the salivary gland chromosomes are *replicated through 10 cycles so that 1000 identical chromatin strands are aligned side by side in exact register*. Microscopic examination of polytene chromosomes reveals dark bands separated by light interbands of differing sizes (Fig. 5.21). Approximately 5000 bands are detectable in the four chromosomes of *Drosophila*. The bands are thought to represent **looped domains** within the chromatin fibres and there is good evidence to support the view that each band is the physical counterpart of the gene, as defined by **complementation** of allelic mutants in genetic studies. Part of this evidence is: (a) the number of bands corresponds fairly closely with estimates of the total number of different proteins; (b) different mutations affecting the same function map on to the same band; and (c) individual bands unfold as a gene starts to synthesise RNA to form **chromosome puffs**, and these puffs appear and disappear in characteristic fashion reflecting the pattern of gene activation during development.

Lampbrush chromosomes have been best characterised in amphibian **oocytes** (immature eggs) that are in extended meiotic prophase which can last for months or even years! The chromosomes are in the form of replicated homologous pairs containing a total of four chromatids. Oocytes are actively building up supplies of RNA and protein required for the subsequent development of the embryo. The lampbrush chromosomes consist of a series of extended loops arising from condensed regions, the **chromomeres** (Fig. 5.22). The loops are actively engaged in RNA synthesis and each loop contains one **transcriptional unit**. Each chromomere and associated loop is thought to correspond with a single gene.

In prokaryotes the single chromosome is organised into a fairly compact structure, the nucleoid, consisting of about 80 per cent DNA by mass, as well as protein and RNA, and this complex is attached to the cell membrane. At least *two of the proteins which bind to the DNA have amino acid*

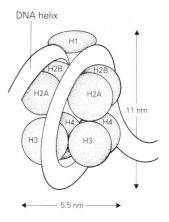

Figure 5.19 Schematic diagram of a nucleosome. About 140 14 bp of DNA double helix is wound around the outside of an octomer ([H2A]$_2$ [H2B]$_2$ [H3]$_2$ and [H4]$_2$) of histones. Histone H1 binds to the outside of the core particle.

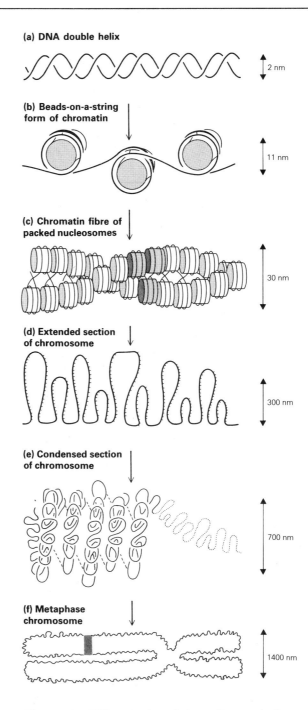

Figure 5.20 Different orders of chromatin organisation.

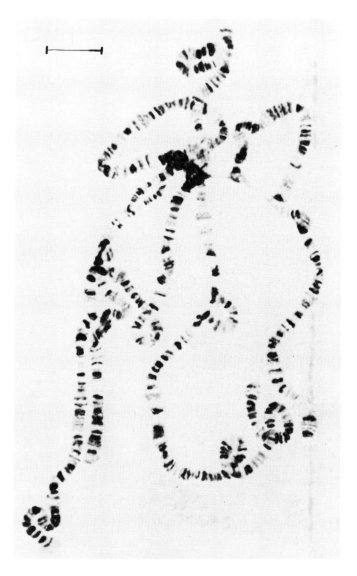

Figure 5.21 Polytene chromosomes of the salivary gland of *Drosophila melanogaster*. Note the highly banded appearance (scale bar 10 μm, courtesy of Dr M. Vald).

compositions resembling those of histones and are thought to organise the DNA into structures directly comparable to the nucleosomes of eukaryotes. A combination of biochemical and biophysical studies has shown that the chromosome of *E. coli* is organised into about 45 loops which radiate out from a dense proteinaceous *scaffold* which is assumed to anchor the DNA (Fig. 5.23). The DNA of the loops is in the so-called **supercoiled** conformation in which the double helix is itself twisted. Supercoiling is a common feature of DNA and arises as a consequence of being covalently closed and circular and/or complexed with protein so that the ends of the DNA cannot rotate freely. Many of the biological properties of DNA which are *dependent on*

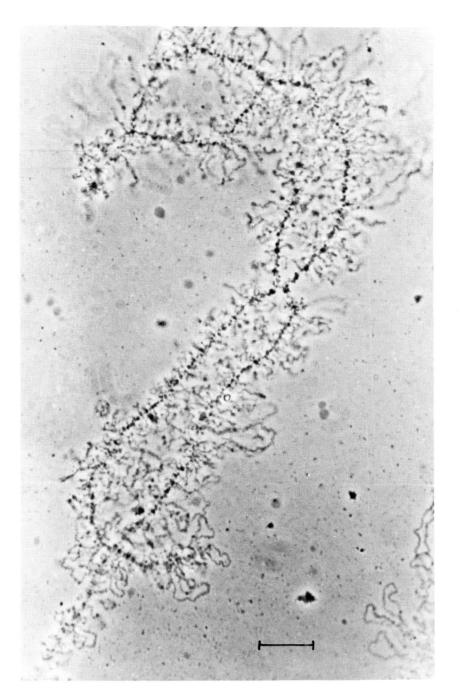

Figure 5.22 Phase contrast micrograph of lampbrush chromosomes of oocyte of the newt (*Triturus cristatus carnifax*). Note the loops of chromatin arising from condensed regions (chromomeres) (scale bar 10 μm, courtesy of Dr M. Vald).

specific interactions with proteins are also dependent on the supercoiled conformation and are lost when the DNA is relaxed by the introduction of a single-strand break. The functional significance of the looped domains of prokaryotic chromosomes is unknown, but they do not represent single units of transcription, in contrast to lampbrush chromosomes.

5.5 Structure of the nucleus

A diagrammatic representation of an interphase nucleus is shown in Figure 5.24. The soluble contents of the nucleus, the **nucleoplasm**, are separated from the cytoplasm by a double membrane system, the **nuclear envelope**. The inner and outer nuclear membranes enclose the **perinuclear space** which is typically 20-40 nm in transverse section. The outer membrane is continuous with the endoplasmic reticulum (ER) and like the ER often has ribosomes

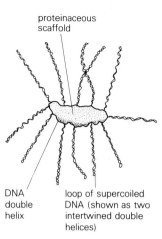

DNA double helix

loop of supercoiled DNA (shown as two intertwined double helices)

Figure 5.23 Schematic representation of the chromosomes of *Escherichia coli*, showing only 12 of the 45 supercoiled loops.

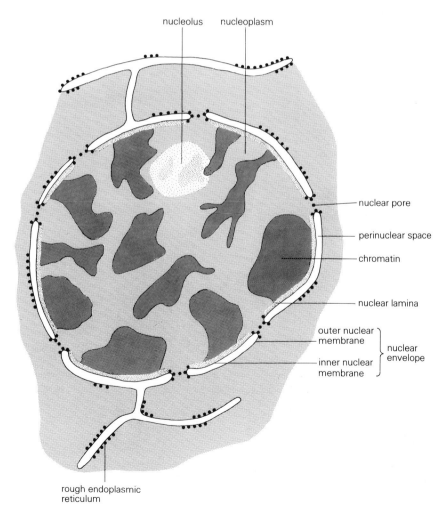

Figure 5.24 Schematic diagram of a cross section of a typical nucleus.

attached to its outer face. The inner and outer membranes are physically continuous because they fuse at intervals and enclose **nuclear pores** surrounded by disc-like **nuclear pore complexes** of about 80 nm internal diameter (Fig. 5.25). Each nuclear pore complex consists of an octagonal array of large protein granules arranged on both surfaces of the envelope. One of the functions of the nuclear envelope and its pores is to regulate the flow in the traffic of RNA and protein molecules in and out of the nucleus. It also serves to preserve an environment in the nucleoplasm in which the plethora of synthetic and degradative reactions going on there can function efficiently (suitable ionic environment, maintenance of pools of precursor nucleotides and histones, etc.).

The double membrane system allows specialisation of the inner and outer membranes for interaction with components of the nucleoplasm and cytoplasm respectively. For example, the cytoplasmic face of the outer membrane can bind ribosomes which are engaged in making proteins which are extruded through the membrane, and discharged into the perinuclear space (see §6.8). Similarly, the inner membrane contains proteins which specifically interact with the underlying fibrous **nuclear lamina**, composed of a set of three proteins. The nuclear lamina has several postulated functions: it is thought to be responsible for the dissolution and re-formation of the nuclear

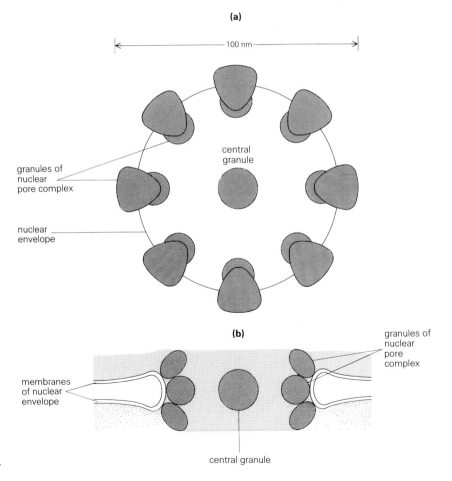

Figure 5.25 (a) Surface view and (b) median vertical section through a nuclear pore complex.

envelope during mitosis and also binds to specific sites on chromatin, thereby positioning the majority of chromosomes around the periphery of the nucleus.

It is assumed that transport of water-soluble materials in and out of the nucleus occurs through nuclear pores, although direct experimental support for this is lacking. Indirect evidence from experiments in which substances are micro-injected into the cytoplasm and their kinetics of equilibration with the nucleoplasm followed, indicates that while small proteins equilibrate rapidly, larger proteins in excess of 60 000 daltons are almost entirely excluded from the nucleus. From this type of study it has been concluded that the nuclear pore complex behaves as if it contained a water-filled cylindrical channel, 9 nm in diameter and 15 nm long. These dimensions pose particular problems for the transport of large **nucleoprotein** complexes such as ribosomes (diameter 15 nm, mass 4.5×10^6 daltons) which are made in the nucleus and exported to the cytoplasm. It is therefore assumed that *such complexes are transported by some active process in which the dimensions of the pores are effectively enlarged.*

The chromatin of the interphase nucleus can be divided into two classes on the basis of its microscopic appearance: condensed chromatin or **hetero-chromatin** and less condensed chromatin or **euchromatin**. Autoradiography (see §1.8.4) of nuclei in cells labelled with ^{3}H-uridine (a precursor of RNA) has shown that *euchromatin is transcriptionally active, whereas hetero-chromatin is inactive.*

The most conspicuous microscopic feature of interphase nuclei are the **nucleoli** – highly granulated structures which are not enclosed by membranes. Each nucleolus is associated with loops of chromatin whose DNA contains the tandemly arranged genes which code for ribosomal RNA (rRNA), as will be discussed in more detail in Section 6.2.1. The granular appearance of the nucleoli results from the accumulation of **ribosomal sub-units** in various stages of assembly. The number and size of the nucleoli depend on the number of **nucleolar organisers** (the region of the chromosomes where the rRNA genes are located) present, the demand by the cell for ribosomes and the phase of the cell cycle. Human diploid nuclei possess eight nucleolar organisers whereas those of the toad *Xenopus laevis* possess only two. The nucleoli disappear as the cell approaches mitosis (the condensed chromatin of metaphase cells being transcriptionally inactive), and reappear in telophase.

The nucleoplasm is not a structureless entity but has a dense fibrillar network, the **nuclear matrix**, running through it. The appearance of the nuclear matrix, whose precise protein composition is unknown, is somewhat similar to that of the cytoskeleton (see §7.3).

5.6   Gene cloning

One of the most exciting and informative areas of current biological research has been the recent development of recombinant DNA techniques. Through the use of these techniques, a wealth of information about genome organisation and expression has been obtained. These advances have been made possible by the development of techniques which allow (a) DNA to be cleaved into defined fragments of manageable size by the use of **restriction endonucleases**; (b) these fragments to be inserted into suitable **cloning vectors** in which the DNA fragments are clonally selected and replicated within their host cell; (c) identification of cloned genes in terms of their encoded polypeptide products; and (d) the rapid sequencing of relatively large stretches of DNA.

5.6.1   RESTRICTION ENDONUCLEASES

In addition to the standard A, T, G and C bases, bacterial DNAs contain varying amounts of bases which have been modified by **methylation**. The precise pattern of methylation is species-specific, and is brought about by a **modification methylase**, the function of which is to protect the cell's DNA from a restriction endonuclease. Restriction endonucleases recognise a sequence of four to six nucleotides in DNA and cleave both strands of the double helix at or near their recognition site. The *in vivo* function of bacterial restriction-modification systems is to protect the cell against foreign invading DNA, such as phage DNA. The host cell's DNA is protected by specific methylation, whereas the foreign DNA may not possess the protective methylation pattern of its host and is thereby degraded. In the order of 150 restriction endonucleases with different recognition/cleavage sites have been identified and about 100 of these are now available commercially. The origin

Table 5.5 Sources and cleavage/recognition sequences of some commonly used restriction endonucleases. The sites of cleavage of the phospodiester linkages are indicated by arrows.

Enzyme	Source	Cleavage/recognition sequence
Hae III	*Haemophilus aegyptius*	$5'$GG$\downarrow$CC$3'$ CC$\uparrow$GG
Msp 1	*Moraxella* spp.	$5'$C$\downarrow$CGG$3'$ GGC$\uparrow$C
Hpa 1	*Haemophilus parainfluenzae*	$5'$GTT$\downarrow$AAC$3'$ CAA$\uparrow$TTG
Eco R1	*Escherichia coli*	$5'$G$\downarrow$AATTC$3'$ CTTAA$\uparrow$G
Pvu 1	*Proteus vulgaris*	$5'$CGAT$\downarrow$CG$3'$ GC$\uparrow$TAGC
Bam H1	*Bacillus amyloliquefaciens*	$5'$G$\downarrow$GATCC$3'$ CCTAG$\uparrow$G
Sal 1	*Streptomyces albus*	$5'$G$\downarrow$TCGAC$3'$ CAGCT$\uparrow$G
Hind III	*Haemophilus influenzae*	$5'$A$\downarrow$AGCTT$3'$ TTCGA$\uparrow$A
Pst 1	*Providencia stuortii*	$5'$CTGCA$\downarrow$G$3'$ G$\uparrow$ACGTC

and sequence specificity of a few of these are shown in Table 5.5. Note that some restriction endonucleases (e.g. Hae III) cut across both DNA strands at the same position to produce DNA fragments with *blunt* ends, whereas others (e.g. Eco R1) make staggered cuts and generate fragments with short complementary *sticky* ends. Through the ability of restriction endonucleases to cleave DNA at specific sites it is possible to construct **restriction endonuclease maps** of a given DNA in which the precise positions of the cleavage sites are located. For example, bacteriophage lambda (λ) which infects *E. coli* has five Eco R1 sites and three Pvu 1 sites. Eco R1 and Pvu 1 digestion will generate six and four fragments respectively, and these can easily be separated from each other by electrophoresis in gels of agarose (see §1.7). By using two or more restriction endonucleases, fragments with overlapping sequences are generated, and it is thus possible to determine the relative order of the fragments in the original molecule. The Eco R1/Pvu 1 map of DNA is shown below:

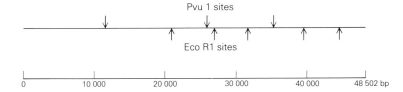

The importance of constructing restriction endonuclease maps is that individual genes can subsequently be accurately located on them.

5.6.2   CLONING VECTORS

Cloning a fragment of DNA allows large amounts of that fragment to be produced in pure form. (A **clone** is defined as a number of identical molecules, cells or individuals all derived from a common ancestor.) Cloning DNA fragments is possible because they can be inserted into cloning vectors – either phages or **plasmids** – which multiply in bacteria or yeast cells such that the 'foreign' sequence is replicated along with the rest of the DNA. An example of cloning using a plasmid will now be considered. Plasmids are relatively small, circular, extrachromosomal DNA molecules capable of **autonomous replication** in their host cell. They differ from viruses in that they do not cause cell lysis, and their DNA is not encapsulated in a protein coat. Some plasmids are present in multiple copies (several hundred) per cell. Many naturally occurring bacterial plasmids carry genes which convey their host cell resistance to antibiotics such as tetracycline and kanomycin, and because of this property constitute a major public health problem. Because of its small size, plasmid DNA can be easily separated from its host's chromosomal DNA and large amounts obtained in pure form. If plasmid DNA is incubated with plasmid-free bacteria in the presence of Ca^{2+}, some of the bacterial cells take up the plasmid, i.e. they become **transformed**. In the case of plasmids specifying antibiotic resistance, transformants are easily recognised by their ability to grow on antibiotic-containing nutrient agar plates (the non-transformants fail to grow). The structure of a plasmid called pBR322, specifically engineered as a cloning vector, is shown in Figure 5.26. Note that this plasmid carries genes which specify resistance to ampicillin (amp) and

Figure 5.26   Plasmid pBR322. The plasmid has several unique restriction endonuclease sites within its antibiotic-resistance genes.

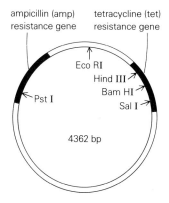

tetracycline (tet) and contains unique restriction endonuclease sites for Bam H1, Sal 1, Hind 111 and Pst 1 within the antibiotic resistance genes. This plasmid is a suitable vector for cloning DNA fragments into any of the above sites. The insertion of foreign DNA into these sites results in a loss of resistance by the host cell to the corresponding antibiotic, since the gene is now interrupted. A hypothetical example of a cloning experiment will be considered using pBR322 as the vector, and mammalian mitochondrial DNA ($\approx$ 18 kbp) as the foreign DNA to be cloned. Both DNAs are first cut with the enzyme Sal 1. This generates a single, linear plasmid molecule and five fragments of mitochondrial DNA (Fig. 5.27). The restricted DNAs are then mixed and incubated in solution at low temperature, such that the DNA fragments will anneal because their ends contain complementary, single-

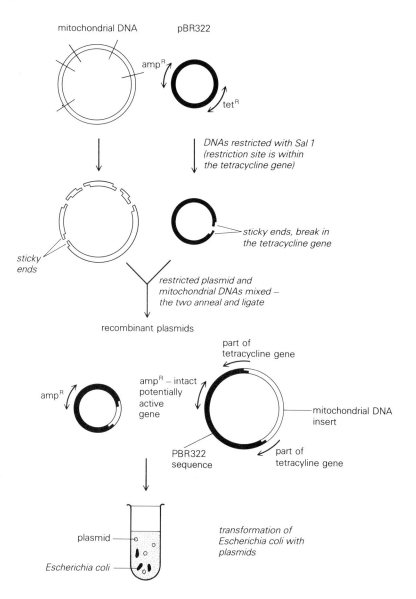

Figure 5.27 Cloning fragments of mitochondrial DNA in Sal 1 site of pBR322.

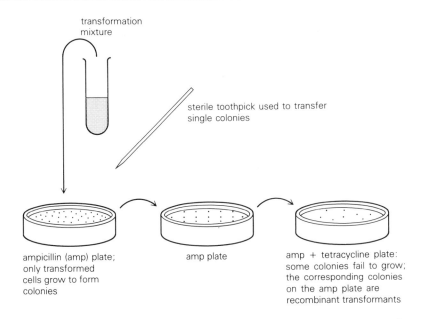

transformation
mixture

sterile toothpick used to transfer
single colonies

ampicillin (amp) plate;
only transformed
cells grow to form
colonies

amp plate

amp + tetracycline plate:
some colonies fail to grow;
the corresponding colonies
on the amp plate are
recombinant transformants

Figure 5.28 Selection of
recombinant transformants.

stranded tails which can form Watson-Crick base pairs with each other. The
precise way in which the DNA fragments anneal is purely a function of their
respective concentrations within the reaction mixture. Thus, some fragments
of mitochondrial DNA will anneal with the plasmid and then re-form circular
molecules (Fig. 5.27). The breaks in the phosphodiester backbone are then
sealed with DNA ligase. The ligated DNA mixture is then used to transform
plasmid-free E. coli. Although only a proportion of the transformed cells will
contain plasmids with inserts of mitochondrial DNA (i.e. **recombinant**
plasmids) these can easily be selected as follows (Fig. 5.28): the transformation mixture is plated onto amp-containing medium. Only transformed cells
will grow and form colonies. Colonies are then individually picked from the
dish (usually with sterile toothpicks) and used to inoculate plates containing
amp and tet. Colonies which grow on both of these plates must have
functional resistance genes for both antibiotics and are therefore non-
recombinant transformants. Colonies which grow on amp medium, but fail to
grow on amp and tet medium are recombinant transformants whose tet-
resistance gene has been inactivated by the insertion of a fragment of
mitochondrial DNA (Fig. 5.27). Such recombinant transformants can be grown
in large quantities and plasmid DNA isolated from them. The identities of the
cloned fragments of mitochondrial DNA can be established simply by
restricting the recombinant plasmids with Sal 1 to cleave out the insert, size
fractionating the inserts and comparing these with the fragments generated by
Sal 1 digestion of mitochondrial DNA. However, for genomes much larger
than mitochondrial DNA, this simple method of identification is not possible
because of the multiplicity of fragments, and it is necessary to resort to the
methods described below. By use of the methods described above it should be
possible to obtain clones of each fragment of mitochondrial DNA. However,
there is a physical limit to the size of fragments which can be cloned using
plasmids (in the case of pBR322, this is about 25 kbp), and clones containing
large fragments tend to be unstable. Derivatives of phage λ are more suitable
cloning vectors than plasmids for large fragments of large genomes.

5.6.3   IDENTIFICATION OF CLONED GENES

By using the cloning techniques described above it is relatively easy to obtain literally millions of clones of fragments of a given genome. The major time-consuming activity is in the identification of clones of sequences of interest, which is analogous to looking for a needle in a haystack. There are a number of different techniques by which this can be achieved, and two commonly used ones are described here.

IDENTIFICATION THROUGH THE USE OF A RADIOACTIVE NUCLEIC ACID PROBE

This technique is mainly applicable to the identification of cloned genes whose products are synthesised in large amounts (for example, α- and β-globin chains in erythrocytes, ovalbumin in the oviduct of laying hens and silk fibroin in the secretory cells of silkworm larvae), and as such their messenger RNAs are abundant and can be isolated in reasonably pure form. A radioactive probe is made from the purified mRNA using the mRNA as a template for **reverse transcriptase** (see §6.6) in the presence of ^{32}P-dNTPs. The product of this reaction is called **complementary DNA** (cDNA). The next step in the identification procedure involves a technique called **nucleic acid**

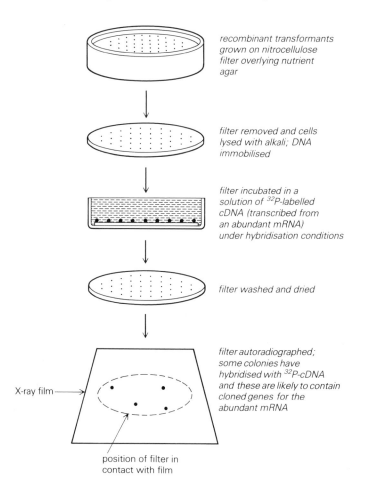

recombinant transformants grown on nitrocellulose filter overlying nutrient agar

filter removed and cells lysed with alkali; DNA immobilised

filter incubated in a solution of ^{32}P-labelled cDNA (transcribed from an abundant mRNA) under hybridisation conditions

filter washed and dried

filter autoradiographed; some colonies have hybridised with ^{32}P-cDNA and these are likely to contain cloned genes for the abundant mRNA

X-ray film

position of filter in contact with film

Figure 5.29   Detection of colonies of recombinant transformants containing cloned genes of interest by colony hybridisation.

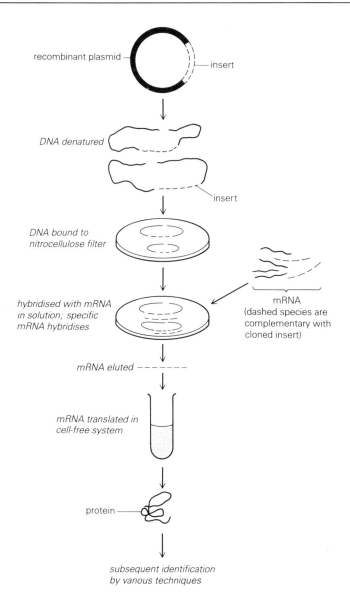

recombinant plasmid

insert

DNA denatured

insert

DNA bound to
nitrocellulose filter

hybridised with mRNA
in solution; specific
mRNA hybridises

mRNA
(dashed species are
complementary with
cloned insert)

mRNA eluted

mRNA translated in
cell-free system

protein

subsequent identification
by various techniques

Figure 5.30 Identification of
cloned genes by hybrid release
translation.

hybridisation. This is done as follows: colonies of the transformed bacteria
are grown on sheets of nitrocellulose filter overlaying nutrient agar plates (Fig.
5.29). The filter bearing the colonies is then treated with strong alkali which
lyses the cells and causes DNA to separate into its complementary strands.
The filter is then washed in buffer of neutral pH and baked at 80 °C; this
causes the single-stranded DNA to become immobilised on the filter,
preventing the complementary strands from re-forming duplex DNA. The
filter is next incubated in a solution of ^{32}P-cDNA at a temperature of about
20 °C lower than the melting point of duplex DNA. These conditions promote
the formation of hybrids between the single-stranded DNA immobilised on
the filter and the cDNA in solution. *DNA hybrids will only form between the
two if they share extensive Watson-Crick sequence complementarity.* The

minimum length of complementary base sequence required to form a stable hybrid under these conditions is about 12 contiguous bases. The filter is then washed to remove non-hybridised cDNA and autoradiographed with X-ray film (see §1.8.4) to locate the radioactive colonies containing DNA-cDNA hybrids. Such colonies should contain clones of the structural gene from which the mRNA used for cDNA synthesis was derived.

IDENTIFICATION OF CLONED GENES BY HYBRID-RELEASE TRANSLATION

This method is often used as a follow-up to the method described above to enable more certain identification of the cloned genes. Colonies of trans-formed cells are grown in liquid media and their plasmid DNAs purified. Individual samples of plasmid DNA are denatured and the single strands immobilised on nitrocellulose filters, as described above. The filters are then incubated with a solution of total cellular RNA (extracted from the same organism as the source of the cloned DNA fragments) under conditions which favour the formation of DNA-RNA hybrids (Fig. 5.30). Again, hybrids will only form if extensive sequence homology exists between the two. Following hybridisations, the filters are washed to remove all unhybridised RNA and then heated in solution to 95-100 °C to melt the hybrid, and release the RNA into solution. This RNA, of complementary sequence to the cloned gene, is next translated into a protein in a cell-free protein synthesising system (see §6.8) containing radioactive amino acids. The radioactive protein made can then be compared in its properties (mobility on gel electrophoresis, immunological properties, peptide fingerprint, etc.) with an authentic cellular protein, and hence establish the identity of the cloned gene.

5.7   DNA sequencing

In recent years, methods have been developed which allow the nucleotide sequence of cloned DNA fragments to be determined simply and quickly. The most important reason for sequencing DNA is to determine the amino acid sequence of the protein encoded by the cloned gene, using the genetic code word dictionary (see §6.7.2) to convert the nucleotide sequence into an amino acid sequence. In theory there are six possible reading frames in DNA (three reading frames on each of the two strands), but it is possible to recognise the coding sequence by the absence of stop codons following the protein synthesis initiation codon (AUG in RNA, TAC in DNA). Such a sequence is called an **open reading frame**.

Two different methods are in common use to sequence DNA. These are the chemical method developed by Maxam and Gilbert and the dideoxy chain termination method developed by Sanger. Both of these methods rely on partial reactions which generate one constant terminus and one variable terminus of the molecule, and both require a detailed restriction endonuclease map of the cloned gene before sequencing can commence.

The fragments produced by these sequencing methods are fractionated according to their size by electrophoresis in a high concentration poly-acrylamide gel (see §1.7), which is able to resolve polynucleotides differing by only one nucleotide in length. The gel is then autoradiographed and the

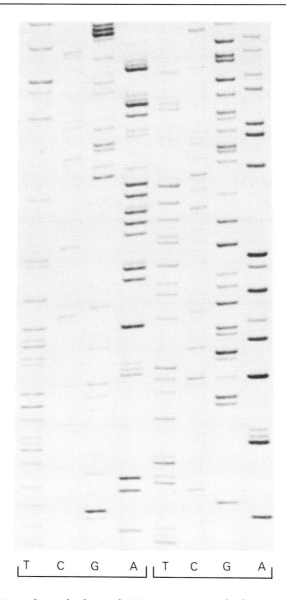

T C G A | T C G A

Figure 5.31 Autoradiograph of part of a DNA sequencing gel. The partial sequences of two genes involved in the regulation of cell division in *Escherichia coli* are shown (courtesy of Dr G. Salmond).

sequence 'read' from the gel. An actual example of a sequencing gel is shown in Figure 5.31. It is possible to read a sequence of up to 200 nucleotides from a single gel. By using these methods, Sanger determined the complete sequence of the 5375 bases of the DNA phage φX174. More than 100 mammalian genes have now been sequenced, and such data are stored in centralised computer facilities.

5.8   Trends in research

Since Chapters 5 and 6 are so closely related the current advances in nucleic acid research have been incorporated into Section 6.9.

References and further reading

BOOKS

Alberts, B., D. Bray, J. Lewis, M. Raff, K. Roberts and J. D. Watson 1983. *Molecular biology of the cell*. New York: Garland.
Lewin, B. 1983. *Genes*. Chichester: John Wiley.
Stryer, L. 1981. *Biochemistry*, 2nd edn. New York: W. H. Freeman.

SELECTED ARTICLES

Cohen, S. N. and J. A. Shapiro 1980. Transposable genetic elements. *Scient. Am.* **242** (Feb.), 36.
Howard-Flanders, P. 1981. Inducible repair of DNA. *Scient. Am.* **245** (Nov.), 56.
Kornberg, A. 1968. The synthesis of DNA. *Scient. Am.* **219** (Oct.), 64.
Kornberg, R. D. and A. Klug 1981. The nucleosome. *Scient. Am.* **244** (Feb.), 48.
Kornberg, A. 1984. DNA replication. *Trends in Biochemical Sciences* (April), 122.

The synthesis of RNA and protein

M. R. Hartley

MODERN VIEWS IN BIOLOGY

6.1   Introduction

In the previous chapter it was shown that genes comprise a linear sequence of the four bases in DNA, and that this sequence of bases somehow determines the order of amino acid residues in the protein it specifies. In this chapter the mechanism by which genes are expressed as proteins and some of the ways gene expression is regulated will be described. Many of the early major discoveries on the mechanism of gene expression were made on bacteria, especially *E. coli*, and their viruses. More recently our knowledge of the equivalent processes in eukaryotes has increased greatly. It is now known that although the general mechanisms involved are *similar in prokaryotes and eukaryotes there are also important differences* between the two groups.

6.1.1   THE CENTRAL DOGMA OF MOLECULAR BIOLOGY

Following their description of the structure of DNA, Watson and Crick pointed out that DNA does not directly serve as a template for the ordering of amino acids for their polymerisation into polypeptides. This follows because in eukaryotes the vast majority of DNA is in chromosomes in the nucleus, whereas protein synthesis takes place on particles called **ribosomes** in the cytoplasm. Therefore an intermediate molecule *must be involved in transferring information from DNA in the nucleus to the cytoplasm*. It was postulated that this intermediate is RNA for the following reasons: (a) RNA is chemically very similar to DNA and could be synthesised on a DNA template such that the RNA molecule is complementary in base sequence to a region of one strand of the DNA double helix; (b) cells actively engaged in protein synthesis are rich in RNA; and (c) short-term labelling of cells with ^{3}H-uridine (a precursor of RNA but not DNA) followed by autoradiography (see §1.8.4) of the cells to determine the intracellular location of labelled RNA revealed that RNA is first detectable in the nucleus, and is subsequently found in the cytoplasm. From this type of observation and reasoning, Crick proposed what has become known as the **central dogma of molecular biology**:

The arrows indicate **flow of information**. Over the last 30 years a vast amount of data has accumulated in support of this scheme.

Before looking at the mechanism of RNA and protein synthesis an important issue will firstly be considered: the way in which different cells or tissues of a given multicellular organism become differentiated to perform their diverse functions. Cells differ from each other because they contain different complements of proteins. For example, a red blood cell which functions in oxygen transport contains large amounts of haemoglobin, whereas a pancreatic acinar cell is rich in the precursors of the digestive enzymes of the pancreatic juice, but contains little or no haemoglobin. Two general mechanisms could account for these facts:

(a)   that the genes expressed are merely a reflection of the genes contained by that cell, and as the organism develops from a zygote there is a progressive loss of genes not required for the maintenance of the

differentiated state – such a situation would require that a full complement of genes are retained by the germ line cells; or
(b) that all the cells of an organism contain a full genetic complement but
     only certain genes are expressed in any differentiated cell type, i.e. gene
     expression is selective in that only certain genes are active in a particular
     cell.

There is limited direct evidence, but a mass of circumstantial evidence, to
show that somatic cells retain all the information required to encode a whole
organism, and are said to be **totipotent**. One of the best pieces of evidence for
this comes from the work of Gurdon on nuclear transplantation in frog's eggs.
The eggs are sufficiently large to enable nuclei removed from other cells to be

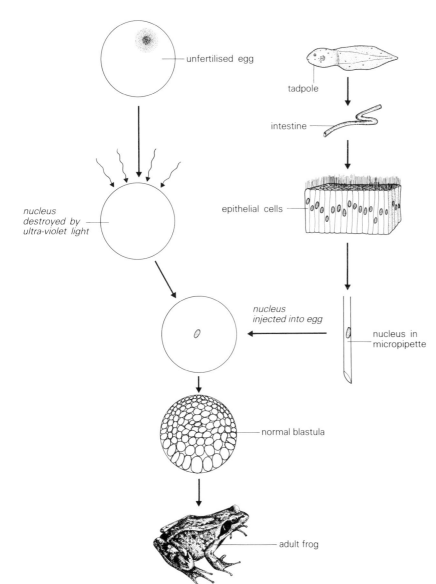

Figure 6.1 Gurdon's experiment showing that the nucleus from an epithelial cell from the gut of a tadpole contains all the information required for the development of an adult frog.

micro-injected into them. In the experiment shown in Figure 6.1 the egg nucleus is first destroyed by irradiation with ultra-violet light, and is replaced with a nucleus from an epithelial cell from the gut of a tadpole. The egg is activated to begin division by the prick of the micropipette used to inject the transplanted nucleus. The activated egg then undergoes the same pattern of development as a normal fertilised egg, and it is possible to recover adult frogs, though with a low success rate. With plants it is possible to regenerate entire individuals from the single cells of differentiated tissue, for example, phloem and leaf parenchyma. There are a few exceptions to the general rule that all the cells of an organism contain the same genes in the same relative proportions. In some invertebrates, chromosome elimination occurs in somatic tissue, and several instances of selective **gene amplification** are known, for example the genes coding for ribosomal RNA (see §6.2.1) in the developing oocytes of amphibia. In these cells there is a particularly high demand for ribosomal RNA and this is met by selectively replicating the genes concerned from the normal (haploid) gene dosage of about 500 copies to some 2×10^6 copies.

6.2   The different classes of cellular RNA

The cells of all eukaryotes and prokaryotes contain three classes of RNA molecules which can be distinguished by structural and functional criteria. These are ribosomal RNA (rRNA), transfer RNA (tRNA) and messenger RNA (mRNA). Both rRNA and tRNA are metabolically stable and are concerned with deciphering the genetic code, but do not code for proteins themselves. On the other hand, mRNA is generally unstable (i.e. is constantly being synthesised and degraded) and carries the nucleotide code which specifies the primary structure of proteins. All of these RNAs are transcribed from DNA templates. The structure and function of these RNAs will now be considered in more detail.

6.2.1   RIBOSOMAL RNA

As the name implies, rRNA is a structural component of ribosomes whose function is to translate the information carried by the nucleotide sequence of mRNA into the amino acid sequence of proteins. Ribosomes are very abundant in cells actively engaged in protein synthesis: for example, *about a quarter of the mass of a rapidly dividing bacterial cell is accounted for by ribosomes*. Each ribosome has a mass of 2.5-4.5 $\times$ 10^6 daltons, and is composed of roughly equal amounts of RNA and proteins. Two classes of ribosomes are found in nature and are usually distinguished by their sedimentation properties in the ultracentrifuge (see §1.5.2). The ribosomes of prokaryotic organisms, chloroplasts and mitochondria have sedimentation coefficients of about 70 S and are smaller than the ribosomes of the cytoplasm of eukaryotes, which have a sedimentation coefficient of about 80 S. Both 70 S and 80 S ribosomes *are composed of two functionally and structurally distinct sub-units* (Fig. 6.2). The ribosome can reversibly dissociate into sub-units; this process is dependent on the Mg^{2+} concentration. The 60 S sub-unit of the 80 S ribosome contains three species of rRNA and about 45 **ribosomal**

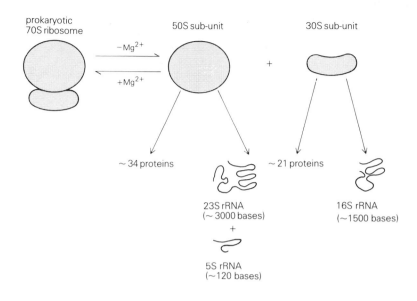

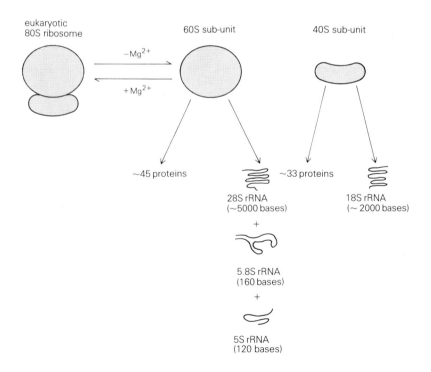

Figure 6.2 The components of 70 S (prokaryotic) and 80 S (eukaryotic) ribosomes.

Figure 6.3 Reconstitution of *Escherichia coli* 30 S ribosomal sub-units. The reconstituted sub-units have functional properties similar to those of native 30 S sub-units.

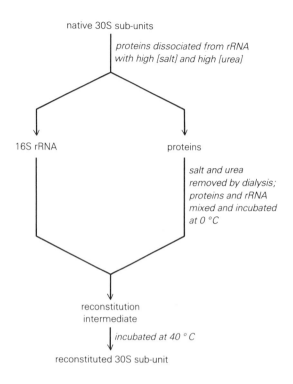

Figure 6.4 Secondary structure of the 3′ region of *Escherichia coli* 16 S rRNA. The single-stranded loops are connected by base-paired double helical stems.

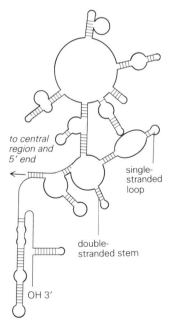

proteins. The 50 S sub-unit of the 70 S ribosome contains two species of rRNA and about 34 ribosomal proteins. The small ribosomal sub-unit (40 S in eukaryotes, 30 S in prokaryotes) contains one species of ribosomal RNA and 21-33 ribosomal proteins. The ribosome of *E. coli* has been studied in great detail and the complete nucleotide sequence of its three rRNAs and most of its proteins have been determined. A variety of physical and biochemical techniques have been used to determine the **topology** (the detailed shape and the positions of the individual protein and rRNA molecules relative to each other) of the ribosome, as well as the functions of the individual components. A particularly useful approach in determining the function of ribosomal proteins is based on the finding that the proteins are 'stripped' off the ribosome when it is suspended in a solution of high salt (e.g. 4 M NH_4Cl) and high urea (e.g. 8 M) concentrations. When the salt and urea are removed from the dissociated proteins and the mixture incubated with rRNA under the appropriate conditions, ribosomal sub-units which are structurally and functionally indistinguishable from native sub-units can be reconstituted (Fig. 6.3). In such **ribosome reconstitution** experiments, a single protein can be omitted from the mixture and the functional consequences of the deficiency determined in assays which measure particular steps in protein synthesis. For example, five proteins of the 50 S sub-unit are required for catalysing the formation of **peptide bonds** in the growing polypeptide chain.

A striking feature of the rRNAs is that they possess a high degree of secondary structure, which arises from intramolecular hydrogen binding between different regions of the same molecule, giving rise to double-stranded **stem** regions and single-stranded **loops** (Fig. 6.4). Such secondary structure is believed to be important for the association of rRNA and proteins.

6.2.2   TRANSFER RNA

In the early 1950s, Francis Crick predicted that small RNA molecules could serve as **adaptors** in that they could react with amino acids and then serve to orientate the amino acids on the mRNA template, the binding to mRNA taking place by means of complementary base-pairing. In this way the genetic code could be read accurately. It is now known that *transfer RNA serves this adaptor role.*

When a cell homogenate is centrifuged at high gravitational forces (e.g. 500 000 × ġ) for several hours, the ribosomes and mRNA pellet to the bottom of the centrifuge tube. The supernatant fraction from such a treatment contains 'soluble' protein and RNA. The majority of this 'soluble' RNA is transfer RNA, as shown by the following evidence: if the supernatant fraction is incubated with radioactively labelled amino acids and ATP, the labelled amino acids become covalently attached to the tRNA molecules. Such **charged** tRNAs then serve as a source of amino acid residues for incorporation into polypeptide chains.

Transfer RNA molecules are classified according to the amino acids with which they become charged. For example, the tRNA that is charged with alanine is called tRNA^{ala}. For reasons that will be explained later in this chapter, the cell contains more than one tRNA species for most amino acids. Different tRNAs which are charged with the same amino acid are termed **isoaccepting** tRNAs. With the development of chromatographic and RNA sequencing techniques in the 1960s, Holley and co-workers determined the primary nucleotide sequence of several tRNAs and about 100 sequences are now known. A striking feature common to all tRNAs is seen from models of structure derived by folding the primary sequence so as to maximise the intramolecular hydrogen bonding and hence produce the most energetically stable conformation. Such studies show that all tRNAs can be accommodated in a **clover leaf** pattern (Fig. 6.5). All tRNA molecules have the following features in common:

(a)   They are single chains containing between 73 and 93 ribonucleotides.

(b)   They contain a high proportion of unusual bases, many of which differ from the standard bases by methylation. Many of these modified bases are **invariant** (i.e. present in all tRNAs), and are believed to be important in maintaining the correct **tertiary structure** (see below, and Fig. 6.6). These modified bases arise by **post-transcriptional modification** of the standard bases.

(c)   The base sequence of the 3' end of all tRNAs is —CCA—OH$^{3'}$. The amino acid is attached to the 3'—OH of the terminal adenosine.

(d)   About half of the nucleotides are base paired to form double helical stems bearing loops of unpaired bases. The loops are termed the **T Ψ C loops** (derived from the sequence ribothymine-pseudouracil-cytosine), the **dihydrouracil loop** (so named because it contains several dihydrouracil residues) and the **anticodon loop**, which interacts with a triplet of complementary bases in mRNA.

The clover leaf model does not tell the whole story about tRNA structure. In 1974 Rich and co-workers deduced the tertiary structure of yeast tRNA^{phe} from X-ray crystallographic studies. The molecule is L-shaped with the dihydrouracil and T Ψ C loops forming the corner of the L and the anticodon loop and amino acid accepting stem at either end of the L (Fig. 6.6). It is believed that all tRNAs adopt this conformation.

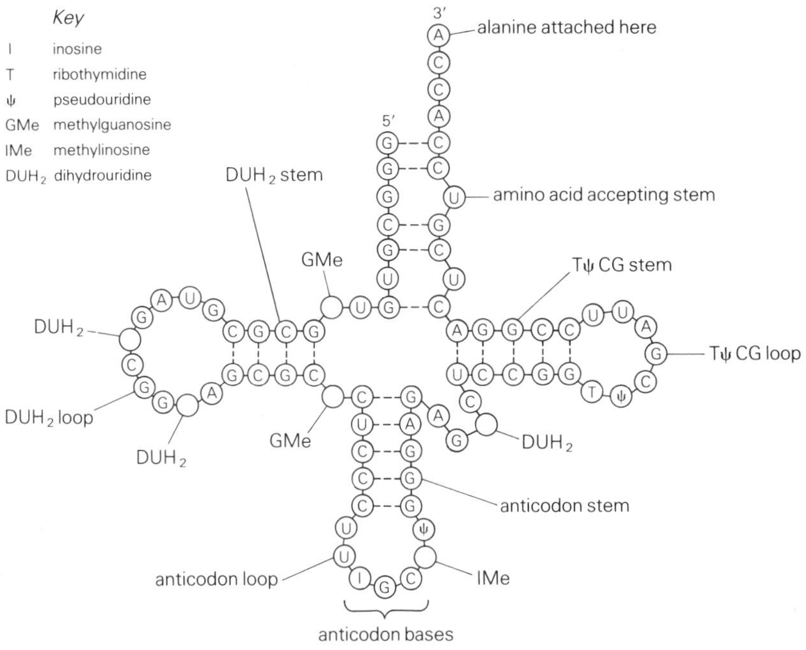

Figure 6.5 Clover leaf structure of yeast tRNA^{ala} showing complete nucleotide sequence. Unusual nucleotides: I, inosine; T, thymine; ψ, pseudouridine; GMe, methylguanosine; IMe, methylinosine; and DUH$_2$, dihydrouridine.

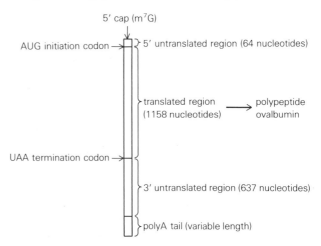

Key
I	inosine
T	ribothymidine
ψ	pseudouridine
GMe	methylguanosine
IMe	methylinosine
DUH$_2$	dihydrouridine

Figure 6.6 Three-dimensional structure of yeast tRNA^{phe}. All tRNAs are believed to adopt a similar conformation.

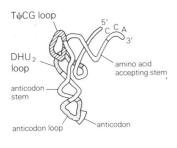

6.2.3 MESSENGER RNA

Most cells are involved in the simultaneous synthesis of several thousand different proteins, so it follows that they contain a similar number of different mRNA molecules. This multiplicity of mRNAs means that *the mRNA population within a cell is heterogeneous in size, unlike rRNA and tRNA populations*. In most eukaryotic and prokaryotic mRNAs not all of the mRNA sequence is translated into proteins. The translated sequence is flanked on either side by **non-translated sequences** which may be several hundred nucleotides in length. Eukaryotic mRNAs also possess two other common features: the base at the 5' end is guanine which has been methylated at the 7 position on the purine ring (the so-called **m⁷G 'cap'**) and the 3' end of the

Figure 6.7 The features of chicken ovalbumin mRNA.

molecule contains a sequence of 100-200 adenylic acid residues (**the poly(A) 'tail'**). Both of these modifications are *post-transcriptional*, and are thought to *confer stability on the mRNA*. Additionally, the cap enhances the binding of ribosomes to the mRNA. The structure of a typical eukaryotic mRNA, that coding for chicken ovalbumin, is shown in Figure 6.7.

6.3   The synthesis of RNA

RNA is transcribed from a DNA template by the action of the enzyme DNA-dependent RNA polymerase, usually abbreviated to RNA polymerase. All RNA polymerases have the following properties:

(a) They require a DNA template, which can either be double- or single-stranded.
(b) All four ribonucleoside triphosphates (ATP, CTP, GTP and UTP) are required for transcription of natural DNA templates.
(c) They require divalent metal ions – either Mg^{2+} or Mn^{2+} – for activity.
(d) *In vivo* only one strand of a particular region of the double helix is transcribed. Transcription is therefore said to be *asymmetric*.
(e) There is no requirement for a primer (cf DNA synthesis, §5.3).

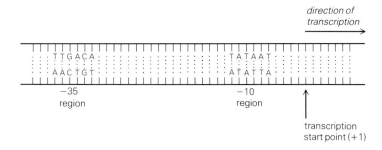

direction of
transcription

-35
region

-10
region

transcription
start point $(+1)$

Figure 6.8   Schematic diagram of a typical prokaryotic promoter. The consensus sequences of the two regions, centred at 10 and 35 bases upstream from the transcription start are shown. The regions are important in RNA polymerase binding.

RNA polymerase catalyses the *initiation* and elongation of RNA chains, i.e.

RNA(n residues) + ribonucleoside triphosphate → RNA(n+1 residues) + PPi

RNA synthesis proceeds by the following steps:

(a) The polymerase binds to a **promoter** region in double-stranded DNA which lies on the 5' side of the gene to be transcribed. In bacteria, promotor sequences are located approximately 10 and 35 bases upstream from the transcriptional start point. Many promoters have been sequenced and have been shown to contain similar sequences, from which it is possible to derive a **consensus sequence** (Fig. 6.8).

   The RNA polymerase of *E. coli* is a large multi sub-unit enzyme with the polypeptide composition shown in Figure 6.9. The whole enzyme complex is termed the **holoenzyme**, whereas enzyme lacking the **sigma sub-unit** is termed the **core enzyme**. During initiation of RNA synthesis the holoenzyme binds to the promoter, this binding being dependent on

Figure 6.9   Schematic picture of RNA polymerase of *Escherichia coli*. The holoenzyme (molecular weight 480 000) is composed of five sub-units of four types $(\alpha_2\beta\beta^1\sigma)$. The σ sigma sub-unit is readily dissociable, forming the $\alpha_2\beta\beta^1$ core enzyme.

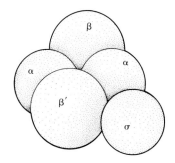

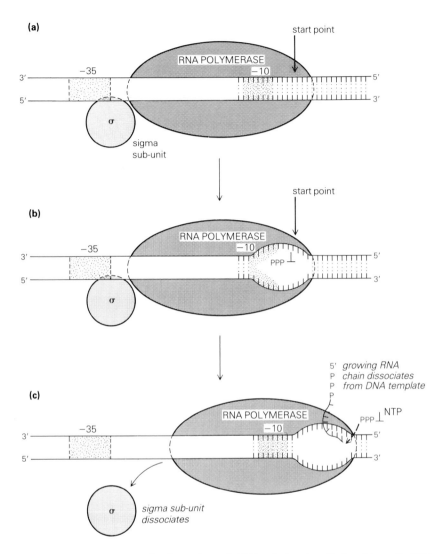

Figure 6.10 Initiation and elongation of RNA chains by *Escherichia coli* RNA polymerase. (a) The polymerase holoenzyme binds to the promoter region of the gene. The −10 and −35 conserved sequences are important in this binding. The sigma unit greatly enhances the affinity of the polymerase for specific binding to these sequences. (b) The polymerase causes local 'melting' of the double helix in the region approximately −12 to +2 bases relative to the transcriptional starting point. The first NTP, complementary to the base at the transcriptional start point, is then bound by the polymerase, and forms a Watson–Crick base pair, as illustrated. (c) RNA chain elongation then commences, and as NTPs are polymerised to the 3′–OH group of the ribose of the growing chain, the transcriptional 'bubble' moves down the transcribed DNA strand in a 3′ → 5′ direction. When about ten nucleotides have been polymerised, the sigma sub-unit dissociates, leaving the core enzyme to complete elongation.

the sigma sub-unit. This sub-unit also causes a local strand separation of the double helix. The first nucleoside triphosphate (NTP), usually ATP or GTP, then binds to the enzyme as directed by the complementary base of the transcriptional start point in the DNA template. A second NTP then binds to the enzyme and the first phosphodiester bond forms between the 3'—OH group of the initiating NTP and the innermost phosphate group of the second NTP (Fig. 6.10).

(b) The elongation of the RNA chain then proceeds. The chain grows in a $5' \rightarrow 3'$ direction and is complementary in base sequence to the transcribed DNA strand (except that uracil replaces thymine). As soon as the RNA chain is about ten bases long, the sigma sub-unit dissociates from the holoenzyme and is available to bind to free core enzyme to initiate a new round of transcription. As transcription proceeds, the hybrid DNA-RNA molecule rapidly dissociates, allowing the newly transcribed region of the chromosome to re-form its native double helical structure. In bacteria ribosomes associate with the new mRNA chains before the latter are complete and protein synthesis begins.

(c) Termination of RNA synthesis is brought about by another 'signal' in the DNA sequence, and a specific chain terminating protein (called **rho** in E. coli) plays a role in this event.

Table 6.1 Properties of eukaryotic RNA polymerases.

Polymerase type	Location	RNA transcripts	Sensitivity to the poison α-amanitin
I	nucleolus	common precursor to 18 S, 5.8 S and 28 S rRNAS	insensitive
II	nucleoplasm	mRNA precursor (heterogeneous nuclear RNA)	highly sensitive
III	nucleoplasm	tRNA precursor and 5 S rRNA	moderately sensitive

Prokaryotes contain a single type of RNA polymerase, though the activity of the enzyme can be modulated by association with different regulatory sub-units. In contrast, eukaryotes possess multiple forms of the RNA polymerase, and different enzymes are responsible for the transcription of the genes for the large rRNAs, mRNAs and tRNAs plus 5 S rRNA. Some characteristics of these enzymes are shown in Table 6.1. The enzymes can be readily distinguished by their different sensitivities to inhibition by $\propto$-amanitin, a cyclic octapeptide from the deadly mushroom Amanita phalloides. As with bacterial transcription, specific signals surround eukaryotic genes which are responsible for the correct initiation and termination of RNA synthesis.

6.4   Transcriptional control in prokaryotes

Many bacteria are found in environments in which the composition of the available nutrients can change markedly. The battery of catabolic enzymes present within a cell growing on a particular medium reflects the cell's need for the efficient utilisation of substrates present in that medium. In general, *bacteria are able to regulate the synthesis of their enzymes in such a way that enzymes are made in significant amounts only when the substrates utilised by those enzymes are available.*

6.4.1   THE LACTOSE OPERON

Escherichia coli cells growing on lactose as their exclusive energy and carbon source contain about 3000 molecules of the enzyme β-galactosidase, whereas cells growing on other carbon sources contain only about three molecules of the enzyme. β-galactosidase splits the disaccharide lactose into galactose and glucose; this conversion is a prerequisite for the utilisation of lactose. Lactose is said to be an **inducer** of β-galactosidase synthesis.

The gene for β-galactosidase in *E. coli* is physically adjacent to two additional genes required for lactose metabolism. These genes code for lactose permease, which is needed for the specific entry of lactose into the cell, and galactoside transacetylase, of uncertain physiological function. A striking feature of these genes is that they show **co-ordinate regulation**. Thus when lactose is added to *E. coli* cells, the amounts of these enzymes increase simultaneously. This arises because the adjacent genes are transcribed into a single **polygenic** mRNA molecule and thus share a common promoter. Such a set of adjacent genes transcribed into a single mRNA and showing co-ordinate regulation is termed an **operon**. The lactose operon of *E. coli* is referred to as the **lac operon**.

It is possible to envisage several mechanisms whereby an inducer could control the rate of enzyme synthesis. It could act at the level of transcription, influencing the rate at which RNA polymerase molecules transcribe the gene in question, or it could influence the rate at which ribosomes translate the mRNA. In fact **transcriptional control** is most commonly encountered. In the example of β-galactosidase *the presence of lactose greatly enhances the rate at which RNA polymerase molecules bind to the promoter.* How is this feat accomplished? The genetic analysis of mutants which are unable to vary the amount of β-galactosidase they produce in response to the presence or absence of lactose was crucial to our understanding of the molecular details. Some mutants synthesise β-galactosidase irrespective of the presence or absence of the inducer, whereas other mutants produce only trace amounts. From these kinds of observations, Jacob and Monod proposed a model to account for gene regulation in bacteria which stands as a landmark in the history of molecular biology. They postulated that a specific **repressor** molecule exists which binds to a stretch of DNA called the **operator** located near to the promoter, thus preventing RNA polymerase from transcribing the structural genes of the operon. The *repressor, now known to be a protein, is coded by a regulatory gene, termed i.* The inducer (lactose) binds to the repressor and in so doing prevents the latter from interacting with the operator gene, allowing the polymerase to transcribe the structural genes. These events are depicted in Figure 6.11, which also shows another aspect of control called **catabolite repression**.

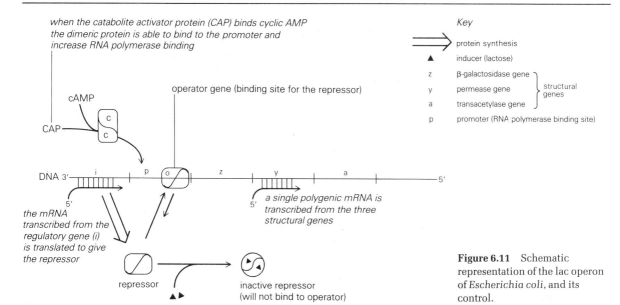

Figure 6.11 Schematic representation of the lac operon of *Escherichia coli*, and its control.

Catabolite repression refers to the inhibitory effect of glucose on the expression of genes which code for enzymes involved in the catabolism of other sugars, such as lactose and arabinose, even when these sugars are present in the growth medium. It would be wasteful to synthesise these enzymes when glucose is available. In the absence of glucose there is a build-up in the intracellular concentration of the regulator cyclic AMP. Cyclic AMP then binds to a protein called **CAP** (catabolite gene activator protein) and the cyclic AMP:CAP complex binds to the lac operon promoter and *creates an additional interaction site for RNA polymerase, thereby increasing the transcription of the structural genes* of the lac operon. In the presence of glucose, the cyclic AMP concentration is lowered and the events described above do not occur.

Thus the lac operon of *E. coli* is under both *positive* and *negative* control. Positive control is mediated by the cyclic AMP:CAP protein and negative control mediated through the binding of lac repressor. Many other examples of operons are known in bacteria and several of these have control circuits similar in principle to those which regulate the lac operon.

In eukaryotes it is usual that enzymes of a particular biochemical pathway, or characteristic of a particular differential state, do not share a contiguous genetic location but are dispersed throughout the genome. Nevertheless, coordinate regulation is still accomplished. The molecular details of the control of transcription in eukaryotes are only just starting to emerge.

6.5   Post-transcriptional modification of RNA

The initial transcription products (termed the primary transcripts) of most genes in both prokaryotes and eukaryotes are *extensively modified by post-transcriptional events before they can assume their ultimate function in the cytoplasm.*

6.5.1   MODIFICATION OF rRNA

The rRNA genes in eukaryotes are present as tandemly arranged repeats clustered in the **nucleolar organiser region** of a particular chromosome. The number of rRNA genes within a nucleolus organiser varies from a few hundred to several thousand depending on the species. The precise arrangement of the rRNA genes is variable, but in all cases the transcribed sequences are separated from each other by **non-transcribed spacers**. The arrangement for *Xenopus laevis* (African clawed toad) is shown in Figure 6.12. The rRNA genes are transcribed by RNA polymerase I to yield a large primary transcript containing the sequences which are to be conserved as 18 S, 5.8 S and 28 S rRNAs, together with non-conserved sequences called **transcribed spacers**. The primary transcript is then processed by a series of cleavages mediated by **exonucleases** (which remove nucleotides from the free ends of RNA chains) and **endonucleases** (which cleave RNA chains internally). The specificity of the nuclease attack is thought to be determined by two factors:

(a) The conserved sequences of the primary transcript are modified by methylation, predominantly at the 2'—O position of the ribose. For example, in human rRNA about 110 methyl groups are introduced at specific positions. Methylation is thought to act as a signal to the nucleases involved in processing, in that *methylated sequences are protected from nuclease attack.*

(b) The nucleolus contains a pool of ribosomal proteins imported from the cytoplasm. These associate with newly synthesised primary transcripts and *may physically prevent nucleases from attacking the conserved regions.*

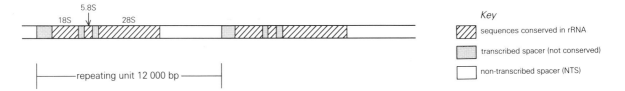

Figure 6.12   Organisation of the ribosomal RNA genes in *Xenopus laevis*.

The primary transcript in human cells is processed as shown in Figure 6.13. Note that the 5 S rRNA (a component of the 60 S ribosomal sub-unit) is not contained within the large transcript. The 5 S rRNA genes are located in different regions of the chromosomes and are transcribed by polymerase III.

6.5.2   MODIFICATION OF mRNA

The production of mRNA in eukaryotes differs considerably from that in prokaryotes in that the primary transcripts generated by RNA polymerase II activity are *extensively processed in eukaryotes before being transported into the cytoplasm.* If eukaryotic cells are incubated for brief periods in the presence of a radioactive RNA precursor, the RNA then extracted and size-fractionated by sucrose density-gradient centrifugation (see §1.5.2), the labelled RNA runs as a broad peak from about 10 S to 60 S (corresponding to RNAs of about 800 to 30 000 bases). Such RNA is referred to as **heterogeneous nuclear RNA** (hnRNA). *The great majority of hnRNA turns over in the*

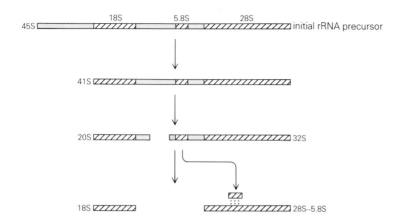

Figure 6.13 Pathway of rRNA maturation in HeLa (human) cells. The primary transcript of the rRNA genes, the 45 S rRNA precursor, is cleaved in the sequence shown to generate the mature forms of 18 S, 5.8 S and 28 S rRNA. The 5.8 S rRNA becomes hydrogen bonded to the 28 S rRNA.

nucleus and never reaches the cytoplasm. The relationship between hnRNA and cytoplasmic mRNA remained an enigma for many years until two key breakthroughs were made in the mid-1970s. The first line of evidence which suggested that hnRNA may be the precursor of mRNA is that both types of molecules often have two types of post-transcriptional modifications in common. These are the m⁷G 'cap' at the 5' ends and the poly(A) 'tails' at the 3' ends of the molecules. The second line of evidence was provided by the surprising finding that most eukaryotic genes encoding proteins are *discontinuous*, in that the protein-coding regions of the genes, called **exons**, are interrupted by non-coding **intervening sequences** or **introns**. The structure of mouse β-globin is shown in Figure 6.14. The three exons of this gene are separated by two introns. It is now known that most, but not all, eukaryotic protein-coding genes are discontinuous, and are referred to as **split genes**. In general, the introns of split genes are longer than the exons. For example, the gene for chicken ovalbumin contains about 7700 base pairs and has 7 introns. The exons account for only 1859 base pairs. The next question we can ask is, at what stage in gene expression are the introns removed to generate functional mRNA molecules? The primary transcript is known to contain the introns present in the gene, so the entire gene is transcribed. The introns are then excised from the primary transcript and the exons are simultaneously *ligated* by a splicing enzyme/ligase complex. Details of how this is achieved are still hazy, *but the mechanism must be very precise because even a one base pair 'slippage' at the intron-exon boundaries would give rise to a protein of entirely different amino acid sequence!* The overall scheme for the synthesis and processing of mouse β-globin mRNA is shown in Figure 6.15. The precise functional and evolutionary significance of split genes is unknown, but there is evidence to suggest that the different exons of some genes code for different functional domains within the protein they encode. A

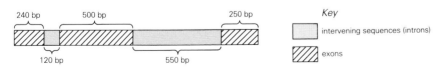

Figure 6.14 Structure of the mouse β-globin gene.

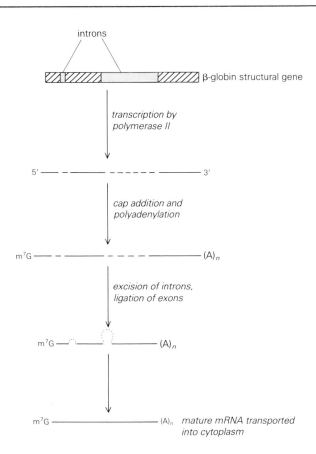

Figure 6.15 Scheme for the formation of β-globin mRNA.

domain is defined as a discrete continuous amino acid sequence which can usually be related to a particular function (see §3.2.2). From such observations it has been proposed that the evolution of many protein-coding genes may have occurred by genetic recombination events which brought together domains previously located on separate genes. Such random events would *occasionally give rise to proteins with 'advantageous' catalytic properties and individuals of this genotype may be at a selective advantage.*

6.6   Reverse transcription

In the central dogma of molecular biology, discussed in §6.1.1, the flow of information from DNA to RNA to protein was considered to be unidirectional. However, in 1970 it was found that certain animal viruses (**retroviruses**), which contain an RNA genome, synthesise double-stranded DNA from their RNA genome. This DNA becomes inserted into the host's genome and is then inherited like any other gene. This finding aroused a lot of interest because many retroviruses are **oncogenic**, i.e. capable of bringing about malignant change. The synthesis of DNA from an RNA template is catalysed by **reverse transcriptase**. This enzyme initially transcribes the viral RNA (using dNTPs as substrates) to yield a DNA-RNA hybrid, and then hydrolyses the RNA of

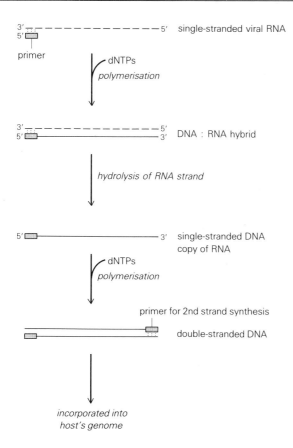

Figure 6.16 Schematic sequence of events whereby reverse transcriptase converts single-stranded retrovirus RNA to double-stranded DNA. The primer required for the synthesis of both DNA strands is believed to be tRNA[trp].

the hybrid by using an associated ribonuclease activity. The same enzyme then synthesises a complementary DNA strand to give double-stranded DNA (Fig. 6.16).

6.7   Genetic code

Before looking at the mechanism of protein synthesis, the details of the way in which the **genetic code** was worked out will be considered. The genetic code describes how proteins containing 20 possible different amino acids are coded by an RNA sequence containing only 4 different nucleotides. A key breakthrough was made by Nirenburg in 1961. He developed a **cell-free protein synthesising system** derived from *E. coli* which was capable of translating **synthetic mRNA**. The cell-free system was made by the procedure shown in Figure 6.17. The bacterial extract was incubated with ATP, GTP, the 20 amino acids (one of which was radioactive in each assay) and the synthetic mRNA. When a synthetic **mRNA homopolymer** containing only uridylic acid residues (—U—U—U—U), called poly(U), was used, it directed the synthesis of a polypeptide containing only phenylalanine residues. Thus the genetic code for *phenylalanine was shown to be an unknown number of uridylic acid residues.*

Figure 6.17 Preparation of a cell-free protein synthesising system from *Escherichia coli*.

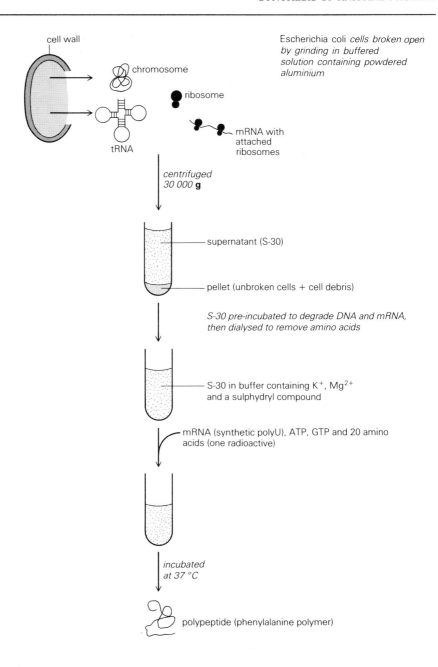

6.7.1 A TRIPLET CODE

Since there are 4 kinds of bases but 20 kinds of amino acids the correspondence cannot be a simple 1:1 relationship. Nor are there sufficient combination of 2 bases ($4^2 = 16$) to account for 20 amino acids. It seems likely then that each amino acid is coded for by a sequence of 3 bases, which would give 64 combinations ($4^3 = 64$) – more than adequate to code for 20 amino acids. The triplet nature of the genetic code was worked out in the 1960s by using a number of different approaches. Some of the best evidence was provided in 1965 by Khorana. By a combination of chemical and enzymatic techniques he was able to synthesise nucleotides of **known repeating sequence** and translate these into polypeptides in a cell-free system similar to that shown in Figure 6.17. For example, the alternating co-polymer UGUGUG . . . codes for a polypeptide consisting of alternating valine and cysteine residues (Val-Cys-Val-Cys . . .). This means that the mRNA is read in groups of odd numbers for the following reasons:

(a)   a triplet codon produces the *alternating* codons

UGU GUG UGU GUG . . .

producing a polypeptide consisting of alternating amino acids, e.g. Val-Cys-Val-Cys;

(b)   a doublet codon would produce the repeat

UG UG UG UG . . .

a homopolymer, or

GU GU GU GU . . .

a different homopolymer;

(c)   a four-letter codon produces the repeat

UGUG UGUG UGUG UGUG . . .

a homopolymer, or

GUGU GUGU GUGU GUGU . . .

a different homopolymer.

Khorana then synthesised copolymers of a 3-nucleotide repeating sequence and was able to establish firmly the triplet nature of the code (try working this out for yourself).

6.7.2   GENETIC DICTIONARY

These experiments could not be used to decide unambiguously which triplet codons coded for which amino acids. These **codon assignments** were most convincingly deduced by means of the **ribosome binding technique**. This makes use of the finding that aminoacyl-tRNA molecules specifically bind to ribosome-mRNA complexes. This binding does not require the presence of a long mRNA molecule; in fact the association of a **trinucleotide** with the ribosome is sufficient to cause aminoacyl-tRNA binding (Fig. 6.18). For example, $^5GUU^3$ caused the binding of val-tRNA and $^5AUU^3$ caused the binding of isoleu-tRNA. All 64 triplets were tested in this system and most gave unambiguous results. The outcome of these and other experiments allows the assignment of 61 out of the possible 64 codons (Table 6.2). The remaining three codons do not code for amino acids but are **polypeptide-chain termination codons**.

Table 6.2 The genetic code.

First position (5' end)	U	C	Second position A	G	Third position (3' end)
U	Phe	Ser	Tyr	Cys	U
	Phe	Ser	Tyr	Cys	C
	Leu	Ser	Stop	Stop	A
	Leu	Ser	Stop	Trp	G
C	Leu	Pro	His	Arg	U
	Leu	Pro	His	Arg	C
	Leu	Pro	Gln	Arg	A
	Leu	Pro	Gln	Arg	G
A	Ile	Thr	Asn	Ser	U
	Ile	Thr	Asn	Ser	C
	Ile	Thr	Lys	Arg	A
	Met	Thr	Lys	Arg	G
G	Val	Ala	Asp	Gly	U
	Val	Ala	Asp	Gly	C
	Val	Ala	Glu	Gly	A
	Val	Ala	Glu	Gly	G

Note: UAA, UAG, and UGA are termination signals. AUG is part of the initiation signal,
in addition to coding for internal methionines.

A number of conclusions can be drawn from the genetic code word dictionary (Table 6.2). Firstly, with the exceptions of the codes for methionine and tryptophan all the other amino acids are encoded by more than one codon, and the genetic code is therefore said to be **degenerate**. Secondly, the same code word dictionary is used by all organisms irrespective of their phylogenies and so it is said to be **universal**. *The only known exception to this rule is found in mitochondrial DNA where a number of varient codon assignments exist (see §2.8). The reasons for this are not known.*

By looking at the genetic code word dictionary we can see why multiple tRNA species exist for most amino acids. The specificity for placing the correct amino acid in a growing polypeptide chain is provided by the Watson-

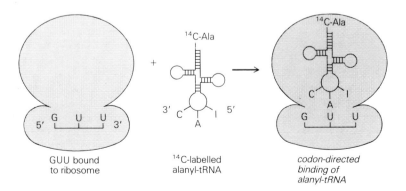

Figure 6.18 Ribosome binding technique for the determination of codon assignments.

GUU bound to ribosome   ¹⁴C-labelled alanyl-tRNA   *codon-directed binding of alanyl-tRNA*

Crick base-pairing interactions between the anticodon of the tRNA and the codon in the mRNA (Fig. 6.18). Because several different codons exist for the majority of amino acids it follows that tRNAs capable of being charged by the same amino acid but differing in the sequence of nucleotides in their anticodons must also exist. Indeed, this is known to be the case.

It might be supposed that a specific tRNA exists for every codon, in which case at least 61 different tRNAs would be present. However, it has been shown that pure tRNAs of known sequence can recognise several different codons. For example, yeast tRNA^{ala} with anticodon bases ^{5}IGC^3 (where I = inosine, shown in Fig. 6.19) could bind to three codons in mRNA, GCU, GCC and GCA. Inosine is frequently found as the 5' base of the anticodon. To explain this type of degeneracy, Crick proposed the **wobble hypothesis**. It states that 'the base at the 5' end of the anticodon is not as spatially confined as the other two, allowing it to form hydrogen bonds with any of several bases located at the 3' end of a codon'. Base-pairing combinations within the wobble concept are shown in Table 6.3. This hypothesis has received direct experimental support.

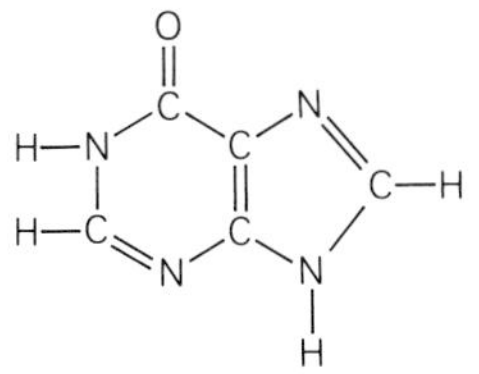

Figure 6.19 The structure of inosine.

6.8 The mechanism of protein synthesis

Protein synthesis is the most complex biochemical transformation which cells perform, and at least 200 different proteins are required for protein synthesis itself. The mechanism of protein synthesis can be divided into the following steps:

(a) The activation of amino acids by ATP.
(b) The attachment of activated amino acids to their own tRNAs to form aminoacyl-tRNAs.
(c) The formation of a complex between ribosomes and mRNA and the codon-directed binding of aminoacyl tRNAs to this complex.
(d) Peptide bond formation.
(e) Polypeptide chain termination and release from the ribosome.

6.8.1 ACTIVATION OF AMINO ACIDS

Before becoming attached to tRNAs the carboxyl group of each free amino acid must be activated by ATP to form an **aminoacyl-adenylate** intermediate (Fig. 6.20a). This reaction, and the subsequent attachment of the aminoacyl moiety of the intermediate to its tRNA, are catalysed by the same enzyme – aminoacyl tRNA synthetase. The aminoacyl-adenylate intermediate remains firmly bound to the surface of its synthetase during this process. The synthetase then transfers the amino acid to the 3'—OH group of the terminal adenylic acid residue of the tRNA to form an aminoacyl tRNA (Figs 6.20b & c). Each amino acid is activated by a specific synthetase, and the attachment of the amino acid to its tRNA is the *only step in protein synthesis in which the identity of the amino acid plays a part*. This was clearly shown in 1962 by Chapeville (Fig. 6.21). Cysteine was attached to its tRNA (tRNA^{cys}) by cysteinyl tRNA synthetase to form cysteinyl tRNA. The attached cysteine residue was then chemically reduced to alanine and the altered molecule

Table 6.3 Allowed base-pairing combinations according to the wobble hypothesis.

5' base of codon	3' base of anticodon
C	G
A	U
U	A or G
G	U, C or A
I	U, C or A

Figure 6.20 Formation of aminoacyl tRNA by aminoacyl tRNA synthetase.

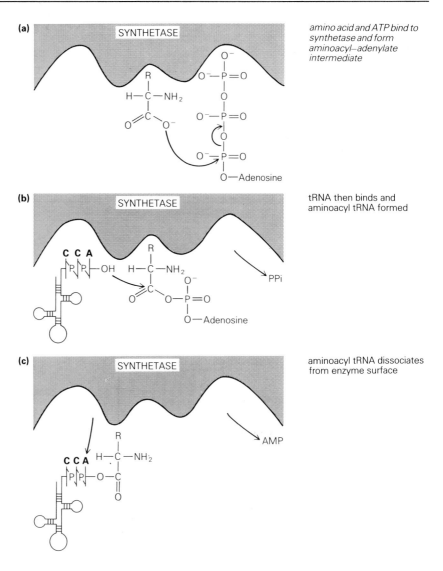

(a) *amino acid and ATP bind to synthetase and form aminoacyl–adenylate intermediate*

(b) *tRNA then binds and aminoacyl tRNA formed*

(c) *aminoacyl tRNA dissociates from enzyme surface*

(alanyl tRNA^{cys}) added to a cell-free protein synthesising system. The resultant polypeptide contained alanine residues in place of the normal cysteine residues.

6.8.2   ROLE OF THE RIBOSOME

Following the aminoacylation of tRNA, the subsequent events in protein synthesis occur on ribosomes. The function of the ribosome is to provide a number of surfaces onto which the other components of protein synthesis – macromolecules including mRNA, aminoacyl tRNA, the growing polypeptide chain and soluble protein 'factors' (see below), as well as small molecules such as GTP – *can specifically bind in a stereochemical fashion which enables the genetic code in mRNA to be read accurately.* The ribosome is able to bind two tRNA molecules simultaneously, one in the **acceptor site** ('A' site)

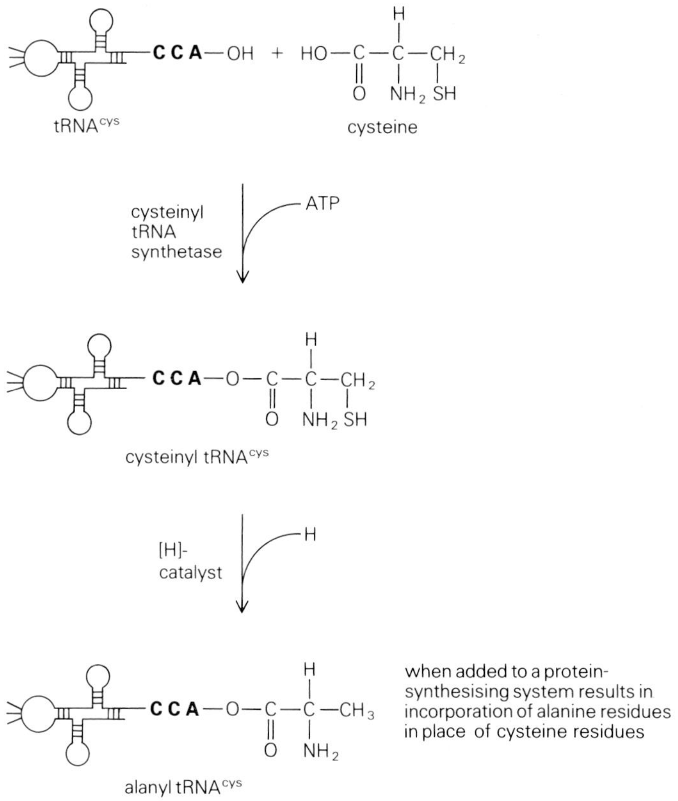

Figure 6.21 The reduction of cysteine attached to its tRNA to alanine attached to the same tRNA.

and one in the **peptidyl site** ('P' site). These sites span across the large and small sub-units of the ribosome (Fig. 6.22). The events on the ribosome can be subdivided into three steps: initiation, elongation and termination. In addition to the part played by the ribosome, each of these steps is also affected by soluble, cytoplasmic protein factors. Several of these factors interact with tRNA, mRNA and ribosomal proteins and functions in bringing these components together in an ordered and specific fashion so that reactions can occur between them.

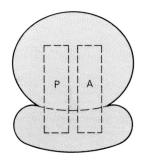

Figure 6.22 Ribosome binding sites for aminoacyl tRNAs. Incoming aminoacyl tRNAs (with the exception of fMet tRNA) enter the A (acceptor) site. During entry into the A site, the P (peptidyl) site is occupied by a tRNA bearing the growing polypeptide chain.

6.8.3 INITIATION OF PROTEIN SYNTHESIS

The events in protein synthesis described below refer to protein synthesis in bacteria, since much more detail is known here relative to eukaryotes. Although the mechanism of protein synthesis on 70 S (prokaryotic) and 80 S (eukaryotic) are thought to be similar there are some significant differences, and some of these will be discussed.

Initiation of protein synthesis begins when the initiator aminoacyl tRNA and GTP bind to free 30 S ribosomal sub-units. In bacteria, the initiator aminoacyl tRNA is always N-formylmethionyl tRNA (fMet tRNA). The structures of methionine and N-formylmethionine are shown in Figure 6.23. Note that N-formylmethionine can only serve as the N-terminal amino acid of a polypeptide because its amino group is blocked (by formylation), so *it cannot engage in peptide bond formation through its amino group.* Next, mRNA joins the 30 S sub-unit-fMet tRNA–GTP complex in such a way that the **initiation codon**, which is nearly always $^5AUG^3$ interacts with the anticodon in fMet tRNA, forming three hydrogen-bonded base pairs (Fig. 6.24). The resulting complex is called the **30 S initiation complex**. Nearly all mRNAs have a non-translated leader sequence at the 5′ ends, and the initiating AUG codon can be 100 nucleotides or so from the 5′ end. Next the 50 S sub-unit joins the 30 S initiation complex to give the 70 S initiation complex. GTP hydrolysis occurs during this step (Fig. 6.24). There are several other features of the initiation process which deserve comment. Firstly, the binding of fMet tRNA, unlike the binding of all other aminoacyl tRNAs, is not **codon directed**, it is in position on the 30 S sub-unit before mRNA bonds onto the ribosome. Secondly, fMet tRNA enters directly into the 'P' site on the ribosome, whereas all other aminoacyl tRNAs enter the 'A' site and are subsequently translocated to the 'P' site. Thirdly, in eukaryotes the initiator methionyl tRNA is not formylated.

Methionine

formylation
blocks peptide
bond formation
at the amino end

N-formylmethionine

Figure 6.23 The structures of methionine and N-formyl-methionine.

Figure 6.24 Steps in the initiation of protein synthesis. Formylmethionyl tRNA, GTP and 30 S ribosomal sub-units interact to form a dissociable complex, which is stabilised by the binding of mRNA to form the 30 S initiation complex. AUG is the initiation codon. A 50 S sub-unit then joins this complex to form the 70 S initiation complex.

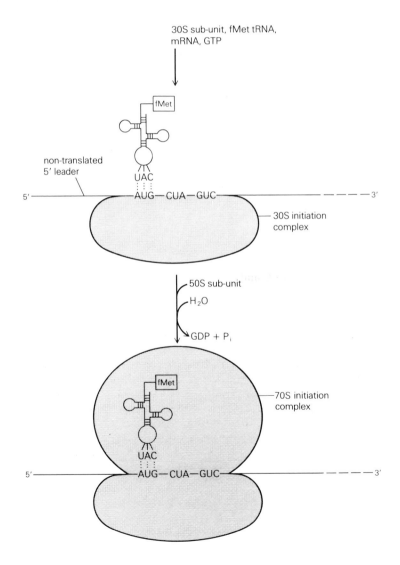

6.8.4   ELONGATION OF THE POLYPEPTIDE CHAIN

The elongation of polypeptide chains can be divided into three steps.

(a) The codon-directed binding of the incoming aminoacyl tRNA to the empty 'A' site of the ribosome. The aminoacyl tRNA inserted depends on the mRNA codon positioned in the 'A' site. GTP is also bound at this stage (Fig. 6.25).

(b) Peptide bond formation in which the amino group of the aminoacyl tRNA rapidly condenses with the carboxyl group of the initiator fmet tRNA in the 'P' site. This reaction is catalysed by peptidyl transferase, an integral component of the 50 S sub-unit.

(c) Translocation in which the peptidyl tRNA in the 'A' site is physically shifted to the 'P' site, displacing the discharged tRNA from the 'P' site. This is accompanied by the ribosome moving along the mRNA in a $5' \rightarrow 3'$ direction by the length of one codon. The energy for this process is provided by GTP hydrolysis (Fig. 6.25).

6.8.5   TERMINATION

Polypeptide chain elongation continues until the ribosome encounters a **termination codon** in the mRNA template. There are no tRNAs with anticodons complementary to the three termination codons. When the ribosome meets a termination codon, the peptidyl transferase functions as a hydrolase, catalysing the transfer of a nascent peptidyl tRNA to water rather than an aminoacyl tRNA (Fig. 6.26). The completed polypeptide chain is thus released and the 70 S ribosome dissociates into its sub-units which leaves the mRNA ready for another round of protein synthesis. In fact, several ribosomes translate an mRNA molecule simultaneously, this structure being called a **polyribosome** or **polysome**.

Protein synthesis is a very accurate process. For example, it has been estimated that the error rate for the replacement of valine by isoleucine (these amino acids differ only in a —CH_3 group on the sidechain) in chicken ovalbumin is only 1:3000.

6.8.6   TRANSLATIONAL AMPLIFICATION

The facts that several ribosomes can translate an mRNA molecule simultaneously and ribosomes can initiate protein synthesis many times on a given mRNA, mean that each mRNA molecule can code for many protein molecules, a phenomenon known as **translational amplification**. An extreme example of this is shown by the production of silk protein (fibroin) by the secretory cells of the silkworm *Bombyx mori*. Each silk fibroin gene, of which there is only one copy per haploid genome, serves as a template for the synthesis of about 10^4 mRNA molecules over a period of several days. Unlike most eukaryotic mRNAs which have half-lives of 2-3 hours, silk fibroin mRNA is stable for several days, and each mRNA is translated about 10^5 times. Thus the single gene is capable of giving rise to 10^9 silk fibroin molecules over a period of about 4 days.

Figure 6.25   Steps in the elongation phase of protein synthesis.

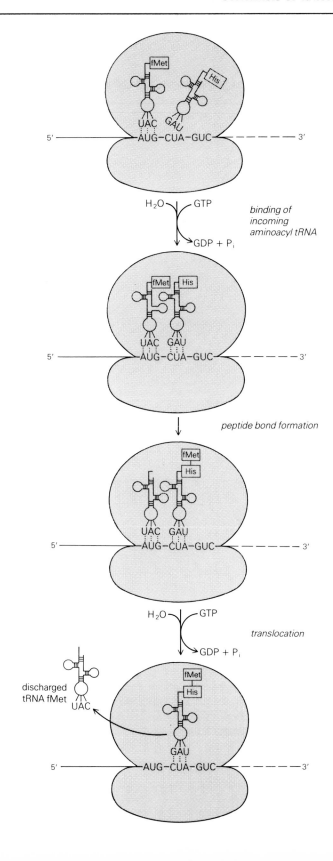

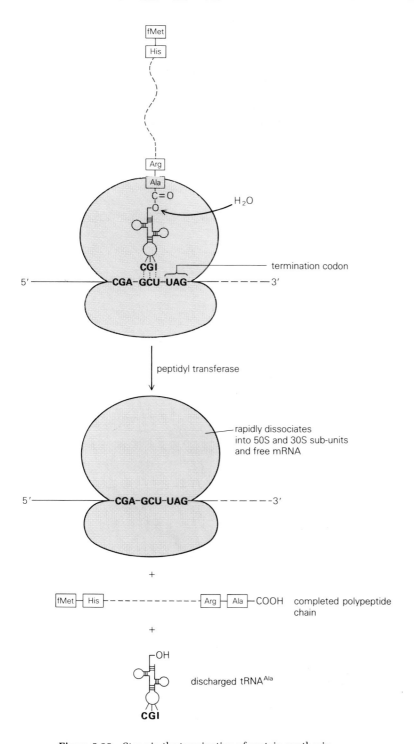

Figure 6.26 Steps in the termination of protein synthesis.

6.8.7   DIFFERENCES BETWEEN PROKARYOTIC AND EUKARYOTIC mRNA

An important difference between prokaryotic and eukaryotic mRNAs is that the former may be **polygenic**, whereas the latter are always functionally **monogenic**. In polygenic mRNAs a single mRNA molecule codes for several proteins (as in operons), each of which is internally initiated on the mRNA template. A well understood example of a polygenic mRNA is that of the small RNA bacteriophages, for example phage MS2, which contain genomes of single-stranded RNA of about 3570 bases. This RNA codes for three proteins: the attachment protein, necessary for adsorption of the phage to the host bacterium and penetration of the phage RNA, the coat protein, which encapsulates the phage RNA, and the synthetase protein, which is involved in the replication of the phage RNA (Fig. 6.27). All three genes have their own ribosome binding sites located near to the initiator codons, and this arrangement means that *the three genes can be translated at differing frequencies depending on the frequency of initiation.* In fact, the coat protein is required in much larger amounts than either the attachment or synthetase proteins and this is reflected in the relative frequency with which ribosomes initiate translation of the coat protein gene relative to the attachment of synthetase protein genes. The frequency of initiation is largely dependent on *the secondary structure of the RNA around the ribosome binding sites,* in that extensive folding of the RNA can block the site in question.

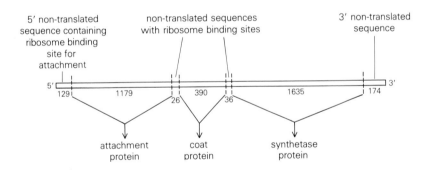

Figure 6.27   The structure of the RNA genome of bacteriophage MS2. The lengths of the three structural genes and non-translated sequences are shown in nucleotides.

Many examples of eukaryotic mRNAs which code for more than one protein are also known. However, in such cases the different proteins are derived from the cleavage by cellular proteases of a large precursor polypeptide and there is only one translational initiation site. An example of this is seen in the translation of poliovirus mRNA (Fig. 6.28) within eukaryotic cells. The limitation of this system is that all the proteins encoded by the mRNA are synthesised in equimolar amounts, even though they may be required in vastly differing amounts.

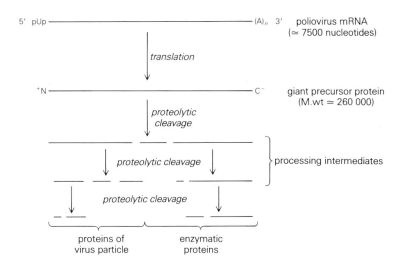

5′ pUp ——————————————————— (A)$_n$ 3′   poliovirus mRNA
                                                        (≃ 7500 nucleotides)

translation

⁺N ————————————————————— C⁻   giant precursor protein
                                                        (M.wt ≃ 260 000)

proteolytic
cleavage

proteolytic cleavage   } processing intermediates

proteolytic cleavage

proteins of   enzymatic
virus particle   proteins

Figure 6.28 The synthesis of poliovirus proteins. The mature forms of the proteins encoded by the viral RNA are generated by multiple proteolytic cleavages of a giant precursor protein.

6.9   Trends in research

6.9.1   TRANSLATIONAL CONTROL OF PROTEIN SYNTHESIS

It is generally accepted that in both prokaryotes and eukaryotes *the most important and widespread point of control of gene expression is at the level of transcription*. Transcriptional control of the lac operon has already been described (see §6.4). However, there are a growing number of instances in which translational control is also known to be an important factor in the production of proteins.

One of the best understood examples of translational control in prokaryotes is seen in the translation of small RNA bacteriophages. As described in Section 6.8.7, the phage RNA codes for three proteins, the attachment protein, coat protein and synthetase sub-unit protein. When the phage RNA enters its host *E. coli* cell, only the coat protein gene ribosome binding site is available for ribosome binding because the other two binding sites are 'closed' owing to the extensive secondary structure of RNA (i.e. intramolecular Watson-Crick base-pairing) in their vicinities (Fig. 6.29). The process of translating the coat protein gene disrupts the secondary structure of the RNA around the synthetase gene ribosome binding site, and allows this gene to be translated. Thus, translation of the synthetase gene is dependent on the simultaneous translation of the coat protein gene (Fig. 6.29). However, the coat protein is required in much larger amounts than the synthetase sub-unit protein, and this is achieved by newly synthesised coat protein sub-units binding to the phage RNA specifically at the synthetase gene ribosome binding site, blocking the further binding of ribosomes. The binding of ribosomes to, and subsequent translation of, the attachment protein gene is controlled in yet another way. When the synthetase sub-unit protein has been synthesised, it combines with three host-encoded sub-units and replicates the positive strand (i.e. mRNA sense strand) into a complementary negative strand. This negative strand then serves as the template for the synthesis of more positive strands for packaging into phage particles.

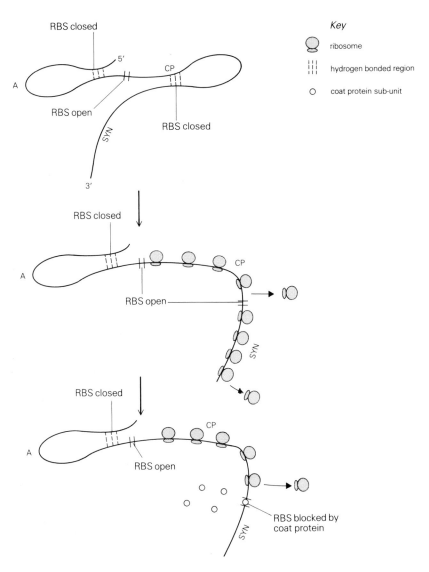

Figure 6.29 Diagrammatic view of the translational controls operating in *Escherichia coli* infected with an RNA phage: A, attachment protein gene; CP, coat protein gene; SYN, synthetase sub-unit gene; RBS, ribosome binding site.

Translation of the attachment protein gene only occurs on incomplete stretches of the positive strands which are in the process of being synthesised from negative strands. This happens because the attachment protein gene ribosome binding site is only accessible to ribosomes just after the commencement of synthesis of the positive strands. The binding site quickly becomes 'closed' as an extensive secondary structure forms around it as the RNA chain is extended further (Fig. 6.30). Thus by a subtle combination of controlling systems which are entirely self-regulating, the phage proteins are synthesised in the ratio of about 20 coat proteins:5 synthetase sub-units:1 attachment protein.

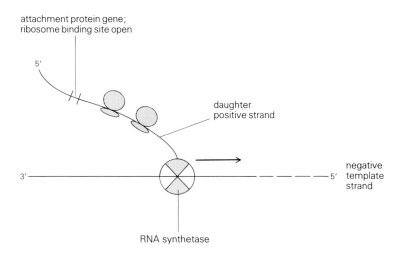

attachment protein gene;
ribosome binding site open

5'

daughter
positive strand

3'   5' negative
                                                    template
                                                    strand

RNA synthetase

Figure 6.30 Translation of the attachment protein gene occurs only on an incomplete positive strand of phage RNA.

Translational control mechanisms in eukaryotes are probably more varied and certainly less well understood than those in prokaryotes. Three general mechanisms have been proposed:

(a) Modulation by one or more of the protein factors, particularly **initiation factors**, involved in protein synthesis.
(b) Modulation through **mRNA stability**. The stabilities of different mRNAs can vary widely, and generally speaking abundant mRNAs are more stable than 'rare' mRNAs. Recently an interesting example of the change in stability of an mRNA has been described. The mRNA for the milk protein caesin, synthesised in mammary glands, becomes much more labile when the hormone prolactin is withdrawn. The molecular mechanism involved is unknown.
(c) Modulation through **mRNA masking**. Messenger RNA *in vivo* is associated with proteins to form messenger ribonuclearprotein (mRNP) particules which may be unavailable for translation. An example of stable, but inactive, mRNA is seen in sea urchin eggs. The unfertilised eggs store mRNA for months in an inactive state. Minutes after fertilisation the rate of protein synthesis increases dramatically, and uses the stored mRNA as template. Again, the molecular details are poorly understood.

The best understood example of translational control in eukaryotes is the control of globin synthesis mediated through haem (the iron-containing prosthetic group of haemoglobin).

Reticulocytes (immature red blood cells) synthesise globin chains at a high rate until haem becomes depleted. The depletion of haem results in the rapid activation of a protein kinase which specifically phosphorylates a protein synthesis initiation factor, called eIF-2, which normally binds GTP and delivers the initiator aminoacy tRNA (Met tRNA$_i$ in eukaryotes, where i indicates initiator) to the 40 S ribosomal sub-unit. Phosphorylation of eIF-2 causes its inactivation, and leads to a halt in the initiation of protein synthesis. Restoration of haem activates a phosphatase which specifically dephosphorylates eIF-2, thereby allowing protein synthesis to recommence (Fig. 6.31).

Figure 6.31 Translational control of globin synthesis in reticulocytes.

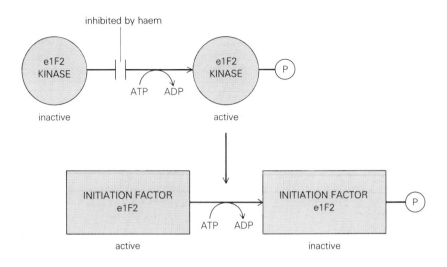

6.10   Applications of recombinant DNA technology

Scientists working in the field of recombinant DNA research were quick to realise the commercial potential of their discoveries. In theory, it should be possible to clone virtually any gene using an appropriate host/vector system and obtain expressions of that gene in the foreign host. The realisation that proteins of pharmaceutical or industrial value could possibly be made more cheaply and in greater quantities than by conventional means, has led to a proliferation of genetic engineering companies, particularly in the USA. A measure of the commercial excitement about the prospects for this new industry was seen during the public launch in 1980 of the US company Genentech. Shares were initially offered at $35 and reached $89 during the first day of trading! This was described by the *Wall Street Journal* as 'one of the most spectacular market debuts in recent history'. Many other applications of recombinant DNA technology have considerable potential. These include the genetic engineering of microbes to perform efficient chemical conversions in the petroleum and mining industries, the detoxification of pollutants, the modification of crop plants by the introduction of genes controlling desirable agronomic traits such as herbicide resistance, disease resistance and nutritional value of seed storage proteins, pre-natal diagnosis of human foetuses for genetic defects and the production of vaccines from cloned viral antigens. The possibilities are almost endless. Three examples will now be considered in more detail.

6.10.1   PRODUCTION OF HUMAN INSULIN

Insulin, needed for the treatment of **diabetes**, is conventionally prepared from the pancreases of slaughtered pigs and cattle. Although sufficient insulin is available from this source to meet present needs, the increasing incidence of diabetes mellitus, which affects 5 per cent of the population of the USA, could mean that demand may outstrip supply. Also, bovine and porcine insulins have slightly different amino acid sequences from that of human insulin, and in some patients are ineffective or may elicit an allergic response. Novo Industry, the largest supplier of insulin in Europe, has recently developed a process that chemically 'humanises' porcine insulin so that it is identical to human insulin, and so obviates the problems discussed above.

Another approach has been adopted by the company Eli Lilly. This involves the production of human insulin in genetically engineered *E. coli*, and *this insulin was the first product of genetically manipulated bacteria to be marketed for human use.* Before considering how this was achieved, the structure and synthesis of insulin will first be considered. Insulin consists of two polypeptide chains, the A chain (21 amino acids) and the B chain (30 amino acids), held together by two interchain disulphide bridges. However, the two chains arise from the cleavage of the storage form of insulin, called **proinsulin**, a process which involves the internal removal of the 33 amino acid C chain (Fig. 6.32). How can bacteria be persuaded to make human insulin? It is not possible simply to insert the human chromosomal gene into a plasmid, transform bacteria and expect insulin synthesis, because the gene contains introns (see §6.5.2) which are not present in bacterial genes and the bacterial cell does not possess the enzymatic machinery for their removal. One approach used in the successful production of insulin in bacteria involved the chemical synthesis of the nucleotide sequences which specify the A and B chains. This was possible because their amino acid sequences were known, so that their DNA sequences could be worked out from the genetic code. Additionally, a 'stop' codon (^{5}TGA^3) was placed at the 3' end of the synthetic genes to ensure polypeptide chain termination and a methionine codon (^{5}ATG^3) placed at the 5' ends for reasons discussed below. The synthetic genes were then inserted separately into the middle of the bacterial gene for the enzyme β-galactosidase carried by the plasmid pBR322, and *E. coli* cells were transformed with the hybrid plasmids. These **fused genes** are expressed from the β-galactosidase gene promoter (see §6.4) and the protein products contain the N-terminal amino acid sequence of β-galactosidase followed by the insulin A or B chain sequences (Fig. 6.33). The A and B

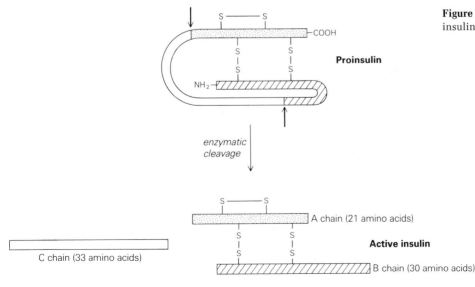

Figure 6.32 The production of insulin from proinsulin.

Proinsulin

enzymatic cleavage

A chain (21 amino acids)

Active insulin

B chain (30 amino acids)

C chain (33 amino acids)

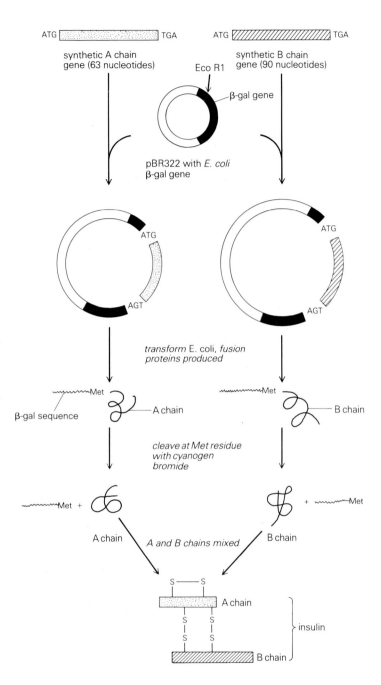

Figure 6.33 Production of human insulin in *Escherichia coli*. From *Recombinant DNA* by J. D. Watson, J. Tooze and D. T. Kurtz. Copyright © 1983 by W. H. Freeman and Company. All rights reserved.

chains of insulin were then liberated from the β-galactosidase amino acid residues by treatment with a chemical (cyanogen bromide) which specifically cleaves polypeptide chains of methionine residues; hence the reason for inserting the methionine codons. Fortunately, insulin A and B chains lack methionine residues, so the insulin chains themselves are not cleaved. The two chains were then purified and mixed together under oxidising conditions which favours disulphide bridge formation. The net result was *pure human insulin*.

A number of other pharmacologically important proteins, including human growth hormone (deficiency of which causes dwarfism) and interferons (antiviral proteins possibly useful in treating certain types of cancer) have also been produced in genetically manipulated bacteria.

6.10.2   GENETIC ENGINEERING IN PLANTS

There are several different techniques involving recombinant DNA which enable plants to be genetically modified. Perhaps the most promising of these makes use of a natural plant host/vector system which is responsible for **crown gall disease**. Crown galls are tumour-like outgrowths found in a wide range of dicotyledonous plants, and consist of masses of proliferating plant cells in which the normal patterns of differentiation have been disturbed. The agent responsible is the bacterium *Agrobacterium tumefaciens*. As well as causing oncogenic symptoms the bacterium diverts the nitrogen and carbon metabolism of the transformed plant cells into synthesising unusual amino acids called **opines** which cannot be utilised by the plant, but which serve as the sole energy, nitrogen and carbon source of the bacterium. How is this transformation achieved? Pathogenic strains of *Agrobacteria* harbour large plasmids, called Ti (tumour-inducing) plasmids. Following entry of the bacterium into the host cell, the plasmids are released, and several copies of a small region, called T-DNA, become integrated into the host cell nuclear genome. The T-DNA carries genes which induce tumour-like growth and opine synthesis in the host cells. By recombinant DNA techniques, it has been possible to derive Ti plasmids in which the genes responsible for tumorous growth have been deleted and into which foreign genes have been inserted. Examples of foreign genes include a seed storage protein gene from french beans and the gene for one sub-unit gene of the photosynthetic CO_2 fixing enzyme ribulose*bis*phosphate carboxylase. These genes are expressed at the protein level, show correct developmental regulation and are transmitted to progeny through the seeds. Thus the technology exists for the insertion and expression of virtually any foreign gene in dicotyledonous plants. It might even be possible to have fields of beans making human insulin! As yet the application of these techniques has not led to the production of plants with superior agronomic traits, but developments should occur within the next few years. Unfortunately, the *Agrobacterium* system is only applicable to dicotyledons, whereas the world's most important crop plants, the cereals, are monocotyledons.

6.10.3   PRENATAL DIAGNOSIS OF GENETIC DISEASES

A major application of gene manipulation is in the diagnosis of genetic diseases in foetuses, so that abortion can be offered in cases where the child is at high risk of being incurably sick. Some 500 genetic diseases which result from recessive mutations in single genes have been identified, and the defective enzyme responsible is known in about 200 of these. These diseases run in families and are inherited in a Mendelian fashion. In heterozygously affected individuals there are usually no symptoms of the particular disease because the normal dominant gene is able to code for sufficient quantities of the protein in question. The problem arises when a couple who are heterozygous for the same gene (carriers) have children. It follows from Mendel's laws that each pregnancy will carry a one in four chance of giving rise to an affected baby with the homozygous recessive gene.

In some instances it is possible to perform prenatal diagnosis for a particular defect where there is a previous history of that defect in either the maternal or paternal family history, in cases where a couple has previously had an affected child, or in individuals of racial/ethnic groups in which the incidence of some disease is much higher than in the population as a whole. Some common genetic diseases, and their frequencies are shown in Table 6.4.

Table 6.4 Some common genetic diseases.

Disease	Defective protein	Approximate incidence among live births
Tay-Sachs	hexominidase A	1 : 3500 Ashkenazi Jews, 1 : 35 000 others
classic haemophilia	clotting factor VIII	1 : 10 000 boys (X-linked)
phenylketonurea	phenylalanine hydroxylase	1 : 15 000
galactosemia	galactoso-l-phosphate uridyl transferase	1 : 40 000
sickle-cell anaemia	β-globin	1 : 400 US Blacks
β-thalassemia	β-globin	1 : 400 in some Mediterranean populations

Foetal diagnosis involves taking a sample of foetal cells (either fibroblasts or blood cells) by amniocentesis, and testing them for the biochemical defect. About 40 different genetic diseases can currently be diagnosed prenatally. However, this rationale has its limitations. The failure to detect a particular enzyme does not necessarily mean that the primary genetic lesion resides in the structural gene for that enzyme (absence could be due to defective

Figure 6.34 Blotting DNA bands from an agarose gel onto a nitrocellulose filter. The strong salt solution is drawn through the gel and elutes the DNA bands. These are deposited on the nitrocellulose filter in contact with the gel, so that the pattern of bands in the gel is replicated on the filter.

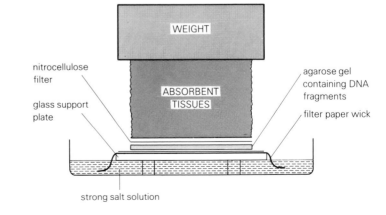

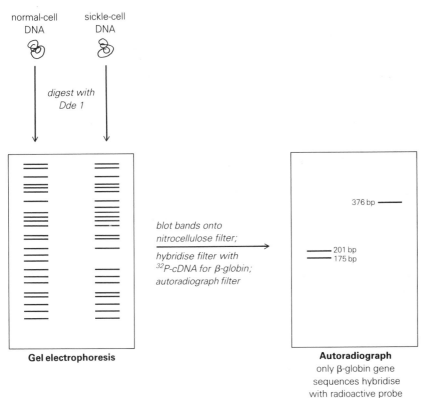

Figure 6.35 Detection of the sickle-cell β-globin gene.

transport, for example) and some diseases cannot be diagnosed in this way because the gene in question is not expressed in the foetus. For example, the defect in sickle-cell anaemia resides in the β-globin chain which is not expressed in the foetus, and which has the haemoglobin sub-unit composition $\alpha_2\gamma_2$. Recombinant DNA techniques allow us to look directly at the genes in question, and so obviate the limitations mentioned above. Sickle-cell anaemia results from a point mutation in the sixth amino acid residue of β-globin, changing the normal glutamic acid residue (DNA code = GAG) for a valine residue (DNA code = GTG). This change in nucleotide sequence eliminates a restriction site for the enzyme Dde 1, cleavage sequence

$$\downarrow$$
$$\text{CTNAG}$$
$$\text{GANTC}$$
$$\uparrow$$

where N = any nucleotide. DNA is extracted from foetal cells, digested with Dde 1 and the fragments separated by agarose gel electrophoresis. The fragments in the gel are then transferred by 'blotting' (Fig. 6.34) onto a nitrocellulose filter and hybridised with a ^{32}P-labelled cloned probe for the β-globin gene. This probe was made by reverse transcribing β-globin mRNA and the complementary, double-stranded DNA cloned in a plasmid vector. Digestion of normal DNA with Dde 1 results in two fragments of the β-globin gene (identified because they hybridise with ^{32}P-cDNA) of 201 and 175 base pairs, whereas sickle-cell DNA gives only one fragment of 376 base pairs (Fig. 6.35).

References and further reading

BOOKS

Alberts, B., D. Bray, J. Lewis, M. Raff, K. Roberts and J. D. Watson 1983. *Molecular biology of the cell*. New York: Garland.

Lewin, B. 1983. *Genes*. Chichester: John Wiley.

Stryer, L. 1981. *Biochemistry*, 2nd edn. New York: W. H. Freeman.

Watson, J. D., J. Tooze and D. Kurtz 1980. *Recombinant DNA*. New York: W. H. Freeman.

SELECTED ARTICLES

Chambon, P. 1981. Split genes. *Scient. Am.* **224** (May), 48.

Chilton, M-D. 1983. A vector for introducing new genes into plants. *Scient. Am.* **248** (June), 36.

Darnell, J. E. 1983. The processing of RNA. *Scient. Am.* **249** (Oct.), 72.

Fiddes, J. C. 1977. The nucleotide sequence of a viral RNA. *Scient. Am.* **237** (Dec.), 54.

French Anderson, W. and E. G. Diacumakos 1981. Genetic engineering in mammalian cells. *Scient. Am.* **245** (July), 60.

Fuchs, F. 1980. Genetic amniocentesis. *Scient. Am.* **242** (June).

Lake, J. A. 1981. The ribosome. *Scient. Am.* **245** (Aug.), 56.

Nomura, M. 1984. The control of ribosome synthesis. *Scient. Am.* **250** (Jan.), 72.

Novik, R. P. 1980. Plasmids. *Scient. Am.* **243** (Dec.), 76.

Rich, A. and S. H. Kim 1978. The three-dimensional structure of transfer RNA. *Scient. Am.* **238** (Jan.), 52.

Cell Biology

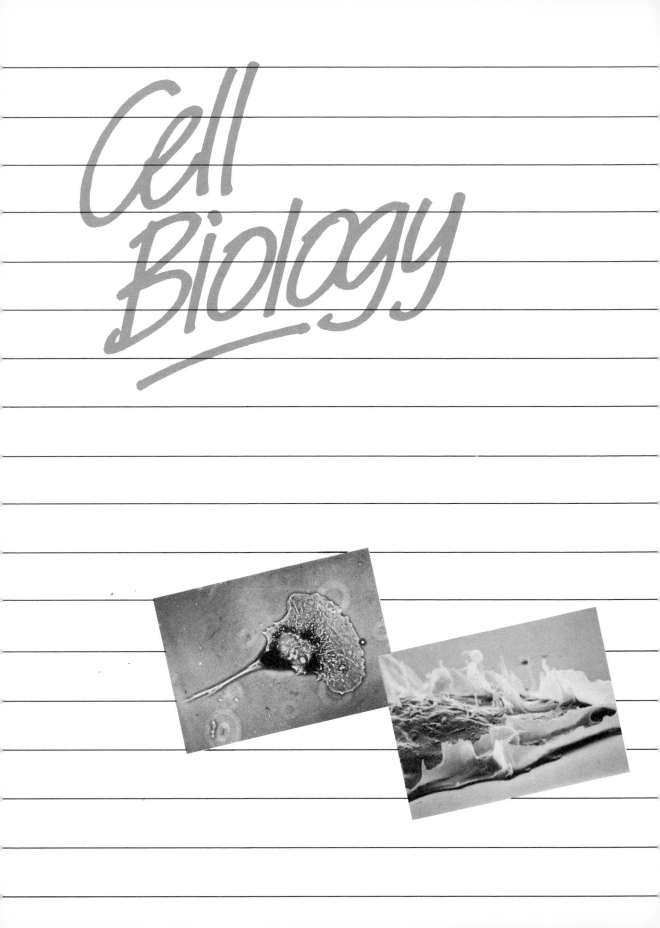

7

Cellular motility

G. E. Jones

MODERN VIEWS IN BIOLOGY

7.1   Microtubules, microfilaments and intermediate filaments

To some degree or other, all cells exhibit motility. The movements that can be seen in different cells vary from the submicroscopic displacements of cell organelles to the gross movements of whole cells seen in developing animal embryos. Whatever the nature of the movement, virtually all eukaryotes use one (or a mixture) of three fibrous structures to execute such rearrangements of their form. Each of the three fibrous structures is built up from certain specific proteins of fairly low molecular weight which have the capacity for rapid assembly or polymerisation under the correct conditions. The three groups of structures are classified on the dimensions seen in electron microscopic preparations as well as according to protein composition, and are called **microtubules**, **microfilaments** and **intermediate filaments**.

Microtubules are hollow cylinders of about 20-25 nm in diameter and vary in length from a few nanometres to tens of micrometres depending upon their location. The best known examples of microtubule-based motility are the *separation of the chromosomes at cell division* and the *movement of cilia and flagella*. Microfilaments are of much smaller diameter than microtubules, about 5-7 nm, are truely fibrous in nature rather than cylindrical and are of variable length. Their role in *striated muscle contraction* has been the most studied function, although research into cell movement has also concentrated recently on the vital role played by microfilaments. The final member of this group of structures, the intermediate filament, was first described in any detail in 1968 and is the most recently discovered of the three. Intermediate filaments consist of filaments with a diameter of about 10 nm, are unusual in that they are made up from tissue-specific classes of proteins, and have largely unknown functions thought to be related to *mechanical integrity of other cell structures*. Because so little is known at present about intermediate filaments, we shall not consider them here, but concentrate largely on the biology of microtubule- and microfilament-based motility. Bacterial flagella will also be described as they are a motile mechanism which is completely unrelated to eukaryotic flagella (despite the name) in both structure and mode of function.

7.2   Microfilament-based motility

The most stable and ordered form of microfilament-based structures is seen in striated muscle. Until very recently our understanding of the structure and functioning of microfilaments was based almost entirely on studies of muscle, and this information was to prove vital to the more recent interest in microfilaments of non-muscle cells. The properties of the microfilament proteins which were characterised through studies on muscle *allowed the same proteins to be identified in virtually every other, non-muscle, cell* and for their role in motility to be rapidly understood. Muscle will therefore serve to introduce the subject of microfilaments and their role in movement.

7.2.1   MUSCLE CONTRACTION

Movements based upon muscle contraction are the best understood of all the mechanisms we shall examine. Muscle cells provide the voluntary movements of the limbs by virtue of an ability to contract rapidly, forcing the skeletal structures, to which they are securely fixed, into different postures. The contraction of muscle is achieved by a very powerful intracellular apparatus, so highly organised *that features of the contraction mechanism can be deduced from its ultrastructure.*

A single voluntary muscle, such as the biceps of the upper arm in man, consists of a large number of elongated muscle fibres. Each muscle fibre has formed during embryological development by the fusion of many separate cells; the muscle fibre is thus a cell containing many nuclei. Most of the mass of a fibre is made up of long strands of 1-2 μm in diameter which extend the whole length of the fibre, called **myofibrils** (Fig. 7.1). Each of these myofibrils consists of a chain of contractile units clearly visible under the electron microscope. Each unit of the chain is termed a **sarcomere**, and is aligned in

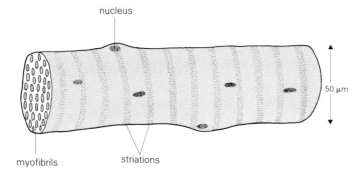

nucleus

50 μm

myofibrils   striations

Figure 7.1 Schematic drawing of a section of skeletal muscle cell (a fibre). The multinucleate cell consists of many myofibrils whose sarcomers are aligned in register. The banding of the fibre represents the repeating dark A bands and light I bands.

register with sarcomeres of adjacent myofibrils, giving the muscle fibre a striated appearance (Fig. 7.2). Sarcomeres are about 2.5 μm long and separated from one another along the length of a myofibril by slender, dark lines known as **Z-lines** or **Z-discs**. Between the Z-lines, the sarcomere is segregated into a series of light and dark bands, called the **I-bands** and **A-bands** respectively. Under the electron microscope the significance of these bands is seen (Fig. 7.3a). Each sarcomere (Fig. 7.3b) is found to contain two sets of parallel and partially overlapping filaments: thick filaments (each about 1.6 μm long and 15 nm in diameter), which extend from one end of the A-band to the other; and thin filaments (each about 1.0 μm long and 8 nm in diameter), extending across the I-bands and partly into the A-band. In relaxed muscle, the thin filaments extend only about halfway along the thick filaments, but it can be shown that each sarcomere shortens as the muscle contracts. In fact, only the I-band decreases in length; the dark A-band remains the same. These observations can be most simply explained if the contraction of a sarcomere *is caused by thick filaments sliding past thin filaments*; there being no change in the length of either type of filament. As each sarcomere shortens, so the whole myofibril (which is made up of thousands of sarcomeres) is shortened. Thus the entire muscle shortens in the lengthwise direction.

This '**sliding filament model**' of muscle contraction was first proposed in 1954. Since that time several lines of study have provided strong support for the concept. At very high magnification it is even possible to see the structural interaction between the thick and thin filaments upon which the sliding filament model depends. Thick filaments can be seen to possess many side arms that extend across the gap to make contact with the thin filaments. It is

Figure 7.2 Electron micrograph of a skeletal muscle fibre (human quadriceps) showing the regular pattern of cross-striations. The many myofibrils run diagonally from left to right (courtesy of Jan Witkowski).

now known that in muscle contraction the thick and thin filaments are pulled past each other by the side arms (more often called cross-bridges) in a manner that we shall examine later. The proteins of thick and thin filaments have now been identified and their biochemistry studied intensively. Most of what we know about these proteins has been acquired from work on muscle or extracts taken from muscle; however, these same proteins are present in non-muscle eukaryotic cells.

7.2.2 MUSCLE FILAMENT PROTEINS

The identity of the thick and thin filament proteins was first established by H. E. Huxley who selectively solubilised both filaments. On the basis of the differential solubility of the two types of filaments, he suggested that *thin filaments consist largely of actin, while thick filaments are made up of myosin.* This work, carried out in the early 1950s, was confirmed by many later studies using other identification techniques.

The structural sub-unit of thin filaments is an almost spherical globular molecule of **actin**, a protein of about 5 nm in diameter, built up from a single, folded amino acid chain. The sub-unit has been sequenced and is known to be

(a)

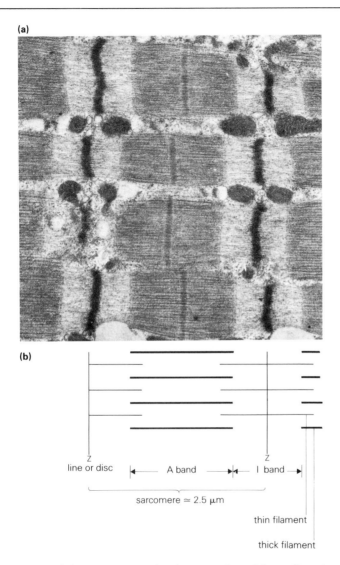

(b)

Z
line or disc

|← A band →|←I band →|

sarcomere ≃ 2.5 μm

thin filament

thick filament

Figure 7.3 (a) Detail from Figure 7.2 showing a portion of four adjacent myofibrils illustrating the structure of the sarcomere (courtesy of Jan Witkowski) (b) Interpretation of the structure of striated muscle, overlapping arrays of thick and thin filaments.

made up from 376 amino acids and has a molecular weight of 41 800. It is conventional to call the actin sub-unit **G-actin** (G standing for globular). The molecule has a binding site for ATP and calcium, and at physiological salt concentrations, *readily polymerises to form filamentous or F-actin*. Hydrolysis of the bound ATP accompanies polymerisation *in vitro*, though it is not absolutely required in muscle cells. F-actin filaments are seen in the electron microscope to consist of two strands twisted into a helix of about 37 nm repeat along its length (Fig. 7.4). The evidence from both X-ray diffraction (see §1.9) and electron microscopy indicates that the thin filaments of striated muscle consist primarily of a single actin helix of this type, containing a total of 300-400 individual G-actin molecules. The thin filaments are composed of more than just actin, however; these components will be discussed later.

Figure 7.4 Drawing of an actin filament to show the helical nature of the polymer with regular repeat of approximately 37 nm.

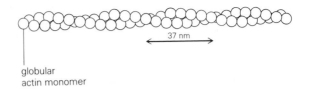

globular
actin monomer

The thick filaments of striated muscle were found to consist of several hundred **myosin** molecules held together by hydrophobic and hydrogen-bonding interactions. Each constituent myosin molecule has a molecular weight of about 500 000 and can be seen in electron micrographs as long, rod-like molecules, each with two globular heads. The myosin molecule can be broken up using detergents or urea into six polypeptide chains. The two largest chains (200 000 daltons each) consist of a pair of identical molecules, each with a globular 'head' and a long 'tail'. In the intact myosin molecule, these two so-called 'heavy chains' align side by side with the two tails coiled

Figure 7.5 Myosin molecule showing the two coiled α-helices of the tail which, together with the globular regions, form the heavy chains. One molecule of each pair of light chains is present on each myosin head.

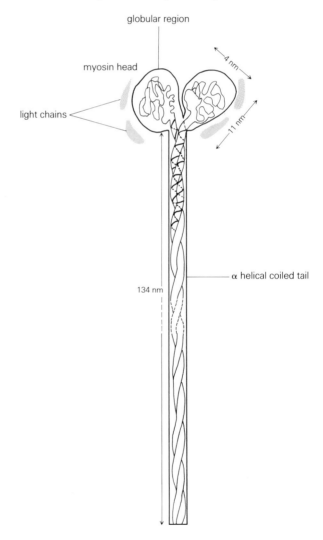

around each other. The four minor polypeptide chains are two pairs of 'light chains'. One molecule of each type of light chain is complexed with each of the two heads of the intact myosin molecule; the whole structure being about 135-140 nm long (Fig. 7.5).

Evidence from a number of laboratories, most notably that of H. E. Huxley, has shown that the most probable arrangement of myosin molecules in thick filaments is one in which the tail units overlap (Fig. 7.6). In the electron microscope, isolated thick filaments appear rough towards their ends while the central section is smooth. This image is interpreted as showing that individual myosin molecules are arranged with their heads concentrated at the two ends of the filament and their tails directed towards the middle. In sectioned muscle, the head units are just visible as the *cross-bridges connecting the thick and thin filaments where the two fibre types overlap.* Each thick filament contains several hundred myosin molecules, packed together in this manner to produce a structure about 15 nm in diameter and a minimum of some 150 nm in length.

7.2.3 ACTIN AND MYOSIN INTERACTIONS

Muscle contraction is a system in which chemical energy is converted into mechanical work. The interaction of actin with myosin is the means by which this is accomplished, with the energy coming from the hydrolysis of ATP to ADP and inorganic phosphate. During muscle contraction, the actin thin filaments slide over the myosin thick filaments, during which the myosin head units attach to the thin filaments. ATP hydrolysis at this stage is a direct consequence of the interaction of myosin head groups with actin. *Myosin has an ATPase activity associated with the head of the molecule which is stimulated by actin.* Under these conditions, each myosin molecule can hydrolyse from 5 to 10 ATP molecules per second. A mass of evidence has now been collected which provides a model for the molecular mechanism of muscle contraction which can be stated as follows. A myosin head carrying ADP and Pi (the products of a previous ATP hydrolysis) moves by random diffusion into close proximity of an actin sub-unit of the thin filament. Myosin then binds to actin, and the ADP and Pi are released from the myosin head. The head tilts by this dissociation and therefore exerts a pull on the rest of the thick filament. This is the *power stroke*, at the end of which a new ATP molecule can bind to the head, resulting in detachment of the myosin head from the actin. The ATP molecule is then hydrolysed by the ATPase activity of the myosin head, relaxing the molecule to its original shape where it is ready for a second cycle (Fig. 7.7).

The idea that ATP binding is necessary for release of the myosin head from actin at the end of the power stroke comes from observations made on muscle deprived of ATP. Such muscle will not relax but enter rigor, a state in which most of the cross-bridges remain firmly attached. It is this effect that causes the state known as **rigor mortis**, the body stiffness that follows death in animals. Examination of muscle in rigor reveals that the myosin heads are locked in a position characteristic of the end of their power stroke: addition of ATP to muscle in rigor releases the myosin heads and allows the muscle to relax. Each myosin molecule is therefore thought to undergo a cyclical change in conformation as it 'walks' along a neighbouring actin filament. The change in conformation causes the myosin to pull against the actin, making it slide along the thick filament. Of the hundreds of myosin molecules that make up a

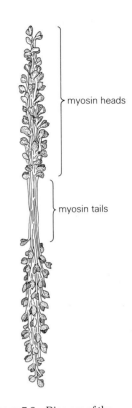

myosin heads

myosin tails

Figure 7.6 Diagram of the arrangement of myosin in the thick filament of striated muscle (not to scale). Tail regions aggregate, leaving their heads to the outside. Note that the polarity of the filament reverses at the bare zone as a result of this arrangement.

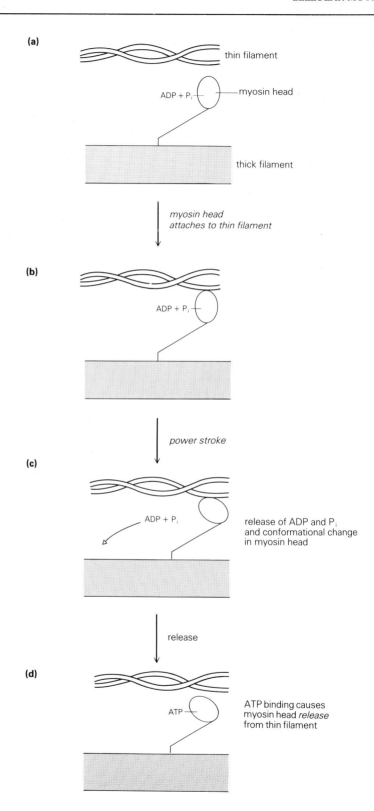

Figure 7.7 Mechanism for the generation of force in striated muscle.

single thick filament, at any one time during contraction some would be attached to actin and others would be in the released state, ready to attach within the next few fractions of a second.

7.2.4 REGULATION OF MUSCLE CONTRACTION

Muscle contraction is under very precise control and, while actin and myosin are the major elements of the contractile apparatus, a number of other muscle proteins are responsible for regulating contraction. A few proteins, such as desmin and α-actinin which are found in the Z-band, are thought to have a structural role in the myofibril. Desmin is thought to help keep the *individual myofibrils of the muscle fibre in register*, so ensuring that Z-bands of adjacent myofibrils are aligned. The very specific localisation of α-actinin at the Z-band, and the fact that this protein binds to the ends of actin filaments in solution, suggest very strongly that it has a role in *anchoring actin filaments in the Z-band*.

The key regulatory proteins of muscle contraction were shown by many groups to be troponin and tropomyosin. It was first discovered that calcium ions were required for contraction of 'native' actin and myosin: a preparation of actin and myosin with the associated muscle proteins. At calcium concentrations below 10^{-8} M, native actin and myosin do not react even when supplied with ATP. Above this level of calcium, the interaction proceeds in the manner already described and cross-bridge formation is seen. Two actin-associated proteins, troponin and tropomyosin, were shown to *mediate this calcium regulation of contraction*. Tropomyosin is a rod-shaped protein with a molecular weight of 35 000. It is arranged as two helical peptides wound around each other in a manner reminiscent of myosin tails and lies in the long pitched grooves on either side of the actin filament. One tropomyosin molecule is long enough to stretch the length of seven actin molecules. Troponin is a complex of three polypeptides called **troponins T**, **I** and **C**. Troponin T has a binding site for tropomyosin and is thought to be responsible for *positioning the complex on the thin filament* via this interaction. Troponin C has four calcium-binding sites but does not react with tropomyosin, though it does bind to troponins T and I.

Since calcium is known to regulate muscle contraction, and since only one polypeptide of the contractile apparatus (troponin C) has any calcium-binding

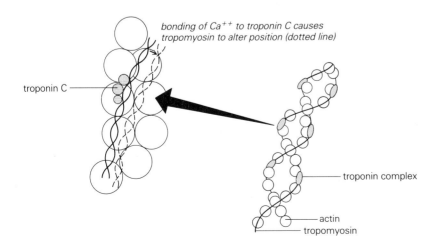

bonding of Ca^{++} to troponin C causes tropomyosin to alter position (dotted line)

troponin C

troponin complex

actin

tropomyosin

Figure 7.8 The proposed arrangement of tropomyosin and troponin in the thin filament. The tropomyosin blocks the sites on actin that bind myosin heads in the contraction phase. When Ca^{2+} binds to troponin C the tropomyosin double helix is displaced, exposing the myosinbinding site on actin.

sites, it is clear that the effect of calcium on the behaviour of this molecule provides the basis upon which contraction is regulated. Detailed structural studies (notably the X-ray diffraction work of H. E. Huxley and his co-workers) have shown the probable mechanism by which troponin and tropomyosin control the interaction of actin and myosin (Fig. 7.8). One troponin complex binds to each molecule of tropomyosin strung out along the F-actin thin filament via troponin T; thus there can only be one calcium-binding troponin C for every seven actin molecules along the filament. In the resting or relaxed state, the tropomyosin molecules lie in or very near the groove of the F-actin double helix; in this position they shield the position on the actin filament that could be occupied by myosin head groups. When the level of calcium is raised, the tropomyosin molecules shift their position very slightly, perhaps falling more completely into the actin grooves. This shift *exposes the myosin-binding sites on the actin molecules, thus permitting attachment of the myosin heads and muscle contraction.* It is supposed (though not yet conclusively demonstrated) that the element responsible for reorientating the tropomyosin molecule is troponin; binding to calcium causes a conformational change in this protein complex, which in turn induces a change in the tropomyosin to which troponin is firmly bound.

The tropomyosin-troponin regulatory system is itself linked to a cellular mechanism that regulates calcium concentration around the myofibrils. Such a system of calcium control is essential for normal muscle function since *very small changes in calcium concentration have dramatic effects on the contractile apparatus.* The initial response to nervous stimulation of striated muscle is the generation of an action potential in the muscle fibre plasma membrane and **sarcoplasmic reticulum** (a system of membranous vesicles surrounding each myofibril). The sarcoplasmic reticulum membranes contain a Ca^{2+}-ATPase pump that, in resting muscle, continually removes calcium ions from the cytoplasm of the myofibrils and stores it inside the vesicles. It is this membrane pump that keeps the calcium concentration of the myofibrils below 1×10^{-8} M, too low for activation of the troponin-tropomyosin complex. The arrival of an action potential at the sarcoplasmic reticulum causes the vesicle membranes to become 'leaky' to calcium and within milliseconds there is a sudden rise in free calcium concentration around the myofibrils to about 1×10^{-5} M. At this concentration the troponin-tropomyosin complex is activated and the muscle contracts. When the nerve impulses cease, the vesicle membranes of the sarcoplasmic reticulum 'close' the channels leaking calcium and the Ca^{2+}-ATPase pump rapidly removes the calcium ions from the myofibrils. When the calcium levels fall to about 1×10^{-8} M the troponin loses its bound calcium and muscle contraction is switched off.

7.2.5 SMOOTH MUSCLE

The description of muscle contraction given above only applies to striated, skeletal muscle and to cardiac muscle, which is responsible for the pumping action of the heart. Cardiac muscle is organised as a tissue in a different manner from skeletal muscle but at the ultrastructural level cardiac muscle has a striated appearance like that of skeletal muscle which reflects a very *similar organisation of actin and myosin filaments* (Fig. 7.9). The third major muscle type present in vertebrates, smooth muscle, which produces slower and longer-lasting contractions in the walls of blood vessels and the intestine, has no striations. Cells of smooth muscle are not multinucleate, and appear as long, thin, tapered bodies containing thick and thin filaments aligned along the long axis of the cell. The filaments are not ordered as strictly as those of skeletal and cardiac muscle, however, and *there is no evidence of myofibrils*.

This difference in structure of smooth muscle is not a reflection of poorer or more primitive construction. Smooth muscle is designed to be able to maintain tension for long periods *while using up some tenfold less ATP than would be required by striated muscle*. This is achieved by two functionally important modifications in smooth muscle myosin. Both actin and myosin of smooth muscle are somewhat different from that of skeletal tissue. The actin differs very slightly in amino acid sequence from skeletal actin, but no functional significance has yet been linked to this. However, while smooth muscle myosin also closely resembles that of skeletal muscle, it has certain other properties. The myosin of smooth muscle is able to interact with actin *only if its light chains are phosphorylated*; when the light chains are de-phosphorylated the muscle relaxes as the myosin can no longer react with actin. A second major difference is related to ATPase activity; in smooth muscle myosin this level is some tenfold lower than myosin of skeletal muscle and there is a more direct calcium regulatory role in this ATPase activity.

Figure 7.9 Electron micrograph of a cardiac muscle cell (myocyte) from the rat to show the similarity of the contractile apparatus (courtesy of N. J. Severs).

Phosphorylation and dephosphorylation of smooth muscle myosin are carried out by specific enzymes. Smooth muscle myosin ATPase is calcium-dependent because the phosphorylating enzyme is regulated by calcium levels. A protein similar to troponin C which is called **calmodulin** (see also §4.5.1) mediates the effect of calcium. Complexing of calmodulin with calcium activates the enzyme responsible for phosphorylating myosin. Calmodulin, which has only fairly recently been discovered, is a protein that *mediates calcium-dependent processes in many cells* and it is now thought that *troponin C is a specialised form of calmodulin that has evolved with the development of skeletal muscle* to provide a more rapid regulation of contraction than is provided in smooth muscle by calmodulin. Smooth muscle myosin shares the property of a requirement for light chain phosphorylation with other myosins; namely those of non-muscle cells that we shall consider next.

7.3   Microfilaments in non-muscle cells

Until about 1965, it was possible to think of cytoplasm as an aqueous phase within which were suspended various organelles such as ribosomes and membranous structures such as mitochondria and in which various meta-bolites were dissolved (see §2.6). Examination of the sometimes highly ordered spatial distribution of organelles in specialised cells led many to suspect that such organisation could not be achieved at random, but that some agent was responsible. The architecture of the cytoplasm is now known to be controlled largely by a system of microfilaments, intermediate filaments and microtubules, together called the **cytoskeleton**.

The framework of the cytoskeleton can be best revealed either by treating cells with non-ionic detergents or by using fluorescently labelled antibodies (see §1.3.3) to the proteins that constitute the cytoskeleton. In the former case, all proteins except those of the fibrous cytoskeleton are extracted; in the latter, specific proteins of the cytoskeleton can be visualised (Fig. 7.10). Once the potential significance of the structures found by these techniques became evident, an area of cell biology using complementary biochemical and morphological techniques rapidly emerged as a unified study of the role of the cytoskeleton in all forms of non-muscle cell movement.

Beginning with the isolation of a protein with the properties of actin from the slime mould *Physarum* in 1966, there have been many demonstrations of the existence of '**muscle proteins**' in non-muscle tissue. Using biochemical isolation procedures as well as antibody labelling techniques, actin has been found in a wide variety of animal and protist cells, algae and the higher plants. Myosin has also been isolated and identified among the protists, fungi, plants and animals. While the actins appear as a major component in most cell protein fractions, myosin has been harder to locate; it occurs in much smaller quantities than the actins and is more variable in structure. Both proteins do seem to be *virtually universally distributed among the eukaryotes* however, and this has led to the concept that actin and myosin probably exist in all eukaryotic cells.

The actin molecules isolated from many non-muscle cells have proved to be closely similar to each other and to muscle actin. Amino acid sequencing has indicated that very few substitutions have occurred in the amino acids of actin during the long evolutionary history of the eukaryotes. Indeed, it has recently become clear that variation is much more likely to be associated with the tissue type of an individual cell within an organism rather than the latter's place in the evolutionary scale.

Myosins isolated from different sources vary much more widely than do the actins. With *only one exception, all myosins possess the same overall shape of muscle myosin molecules*, with two globular head units at the end of a rigid tail section. The exception is a myosin with no tail that has been found in a protist, *Acanthamoeba*; this organism also contains the more familiar myosin. All the myosins bind actin with a resulting increase in their ATPase activity, and they have properties more like the myosin of smooth muscle than striated muscle. Activation of non-muscle myosin, for example, depends upon the phosphorylation of the myosin light chains. As with smooth muscle myosin, phosphorylation (which is catalysed by a myosin kinase) is stimulated by calcium, so that the ability of the myosin to interact with actin is *altered by small changes in the concentration of free calcium in the cytoplasm*. Such changes in calcium levels commonly occur in response to extracellular signals.

The actin and myosin molecules detected in non-muscle cells prove to be responsible for a variety of mechanical and contractile activities in cells. An example of a structural or mechanical role can be seen in the **microvilli** of epithelial cells. In areas where cellular function requires an increase in surface area, such as the intestine, finger-like extensions can be seen to cover the exposed surface of the cell. Each extension is a microvillus, usually about 1 μm in length and 0.1 μm in diameter, enclosed by plasma membrane. The core of the microvillus contains some 40 actin filaments running in parallel along its length. The tip of the microvillus contains amorphous material in which the actin filaments terminate. Evidence on the orientation of the actin suggests that the *amorphous region is equivalent to the Z-band of striated muscle*. At the base of the microvillus, the filaments extend into a network of

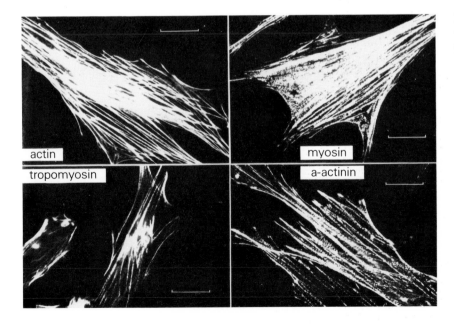

Figure 7.10 Various protein components of the cytoskeleton stained with fluorescently labelled antibodies to muscle proteins: bars = 25 μm (courtesy of J. Heath).

actin filaments called the **terminal web**, which, as it also contains myosin, may create the tension required to keep the microvillus in an upright position. Some of the most intensively studied systems using cytoplasmic filaments are those in which a contractile function is paramount. These include cytoplasmic streaming in plant cells, amoeboid locomotion and the movement of tissue cells, all of which as we shall see depend on a very similar interaction between actin and myosin.

7.3.1   CYTOPLASMIC STREAMING IN PLANTS

Plant cells are generally very large in comparison to animal cells, being commonly over 100 µm long. Such sizes are massive on the molecular scale and important physical processes such as diffusion of molecules become very inefficient. Large plant cells display an extensive cytoplasmic streaming which moves material around the cells very rapidly and must serve to overcome the problems inherent in large size. The massive cells of the green algae *Nitella* and *Chara* provide the most dramatic demonstration of cytoplasmic streaming. In these cells, a continuous ribbon of cytoplasm streams in a helical path sandwiched between a subcortical layer of chloroplasts and a large central vacuole. The flow of cytoplasm is unidirectional and speeds of nearly 100 µm per second have been recorded. Under the light microscope, thin fibrils can be seen in a zone separating the static subcortical layer and the moving cytoplasm. Electron microscopy reveals that each of these fibrils consists of a bundle of actin filaments aligned with the same polarity, opposite to the direction of cytoplasmic streaming. These observations are consistent with the proposal that *streaming is generated by sliding between actin microfilaments, which are embedded in the immobile subcortical layer, and myosin molecules attached to elements in the moving cytoplasm* (Fig. 7.11). Detection of the myosin in the algae has proved difficult but has been reported recently, although our knowledge of plant myosins remains generally very poor. Actin has been found in a variety of plant cells, though little is known yet of their detailed amino acid content

Figure 7.11   The giant algal cell *Nitella* appears to use actin–myosin interactions generating cytoplasmic streaming. The diagram shows the arrangement of the superficial layers of such a cell. Myosin is thought to cross-link elements in the moving cytoplasm with the actin filaments.

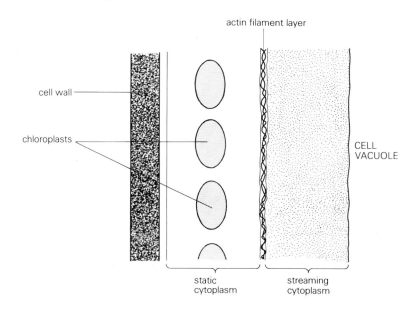

and sequence. Nevertheless, cytoplasmic streaming in plants is thought to be governed by actin filament/myosin interactions which are similar to those seen in non-muscle cells described in the next section.

7.3.2 CELL LOCOMOTION

Plant cells remain fixed within their walls and do not migrate. Animal cells and those of many protists, such as the group to which amoebas belong, can and do move extensively under appropriate conditions. Cell locomotion depends upon a complex interaction of many components of the cytoskeleton which will be examined in more depth towards the end of this chapter (see §7.4.3). One of the components of locomotion is a layer of actin filaments that lie just under the plasma membrane which has been studied most thoroughly in the protists such as amoeba. Amoeba moves by extending and retracting short, thick processes called **pseudopods**. In contrast, the locomotion of most vertebrate cells is accompanied by the formation of thin sheet-like extensions called **lamellipodia**. Despite this gross difference between amoeboid and tissue cell movement, *both are inhibited by drugs which specifically disable actin filaments*. This suggests that both forms of locomotion depend on the functioning of actin filaments.

7.3.3 AMOEBOID LOCOMOTION

The light microscope can be used to distinguish two regions of cytoplasm in amoeba, a central zone of fluid endoplasm and a surrounding, more transparent, gel-like ectoplasm. During pseudopod extension, the endoplasm can be seen to stream in the direction of extension, congealing into the more solid ectoplasm at its tip. Towards the rear of the cell, ectoplasm appears to transform into endoplasm and stream towards the pseudopod. These changes from the viscous ectoplasm to fluid endoplasm are known as **gel-sol transitions**. The exact nature of amoeboid locomotion is still controversial; investigators disagree as to whether contraction of the ectoplasm or cytoplasmic streaming is responsible for the flow of endoplasm to the extending pseudopod. According to some studies, the motive force is generated by contraction of parts of the ectoplasm; this forces the more liquid endoplasm passively towards the tip of the amoeba. Others maintain that the

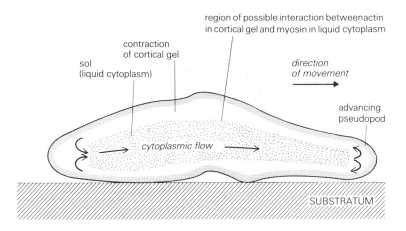

Figure 7.12 Model of amoeboid locomotion. Cytoplasm moves through the cell interior by contraction of the cortical gel and/or by interaction between microfilaments in the gel and myosinlinked structures in the interior sol.

movement of endoplasm occurs actively by a form of cytoplasmic streaming very similar to that described earlier for plant cells. At present there is no reason to discard either idea, and it is entirely possible that *both* mechanisms are at work (Fig. 7.12).

Whatever the case, a good deal of evidence suggests that the movement is actin-based. Mention has been made of the inhibitory effect of actin-binding drugs, and it is now possible to extract the cytoplasm of amoeba and to perform experiments on these crude extracts. The extracts contain actin in the filamentous state and are associated with a number of actin-binding proteins. It has been discovered that some of these actin-binding proteins, notably α-actinin and filamin, can link together actin filaments so as to produce a random network. This has the effect of converting the viscous fluid of the original preparation into a solid gel. Other actin-binding proteins cause a reversal of the physical state from gel to sol; in conditions of high calcium concentration, a protein named 'gelsolin' breaks up the network of actin and filamin. Gelsolin binds so strongly to actin in the presence of calcium that the gelsolin molecules are able to insert themselves between the sub-units of actin in the filament, causing them to disassemble.

A mixture of actin filaments and associated actin-binding proteins like filamin and gelsolin is capable of undergoing gel-sol transitions. However, this artificial mixture will not display the contraction seen in crude cell extracts. *The presence of myosin is crucial to this behaviour.* Myosin is present in very small amounts in the cytoplasm of amoeba; about 2 per cent of the total protein compared to 20 per cent for actin. It seems likely that myosin does not organise into characteristic muscle-like thick filaments, but may exist as aggregates too small to visualise. Even bipolar aggregates of myosin molecules could form *functional small sarcomeres by pulling one set of actin filaments against another as long as the actin filaments were arranged with opposite polarity to one another.* In summary, we are still a long way from a complete understanding of the molecular basis of amoeboid locomotion. We do now have enough information at least to construct a model of the basis of gel-sol transformations which form the basis of this phenomenon. Interaction of actin filaments with cross-linking proteins such as filamin produce a gel. Gelsolin will disrupt the gel following calcium stimulation, and convert the cytoplasm to a sol. At the same time, the same rise in calcium will activate myosin molecules, leading to a pulling of actin filaments against one another, generating fluid streaming.

7.3.4   TISSUE CELL MOVEMENT

In adults, tissue cell movement is normally restricted to a few cell types such as **leucocytes**. Only in certain rare incidents will active tissue cell migration be observed, the one common example of this being the *movement of skin fibroblasts into a superficial wound from surrounding undamaged tissue.* There are also some diseased states in which cell movement can be seen; this is particularly evident in certain cancers where *proliferating cells of the primary tumour can spread rapidly through neighbouring tissues.* In general, cell movement is normally seen only during stages in the development of the embryo. This is most dramatically demonstrated at the stage of **gastrulation** in vertebrates, when large-scale migrations of cells will occur as the whole embryo is reorganised into the three primary tissue layers prior to organ formation.

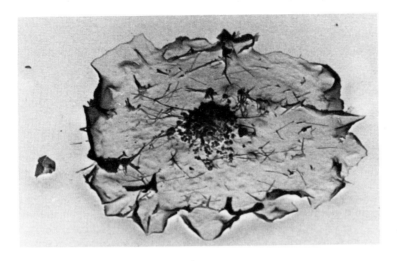

Figure 7.13 Scanning electron micrograph of an embryonic fibroblast from a chick heart in the early stages of spreading on a substrate. The small surface projections (filopodia) can be seen over the nuclear area of the cell. These disappear as the thin sheets of lamellae spread from the cell body (courtesy of J. Heath).

Largely because of the fundamental role of cell movement in embryonic development and its possible involvement in the spread of some cancers, there has been an intensive study into the mechanism of cell locomotion. Because living embryos are complex mixtures of cell types, with individual cells frequently difficult to distinguish, most studies have been carried out in tissue culture. Cells are removed from an organ and grown on a two-dimensional surface of culture vessels made from glass or plastic. The cells attach and move on these surfaces provided that they are given an aqueous medium containing nutrients. While such an environment is very different from the three-dimensional nature of an embryo, such tissue culture systems provide many advantages to investigators and has led to a more rapid characterisation of the locomotion of cells than would have been possible otherwise.

When isolated tissue cells are first suspended in culture medium, they assume an approximately spherical shape. The surface is covered with small

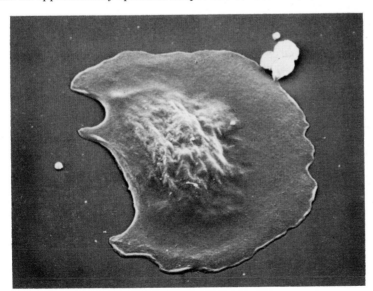

Figure 7.14 Scanning electron micrograph of a chick heart fibroblast to show the 'fried egg' morphology of a well spread cell prior to the development of polarity (courtesy of J. Heath).

(a)

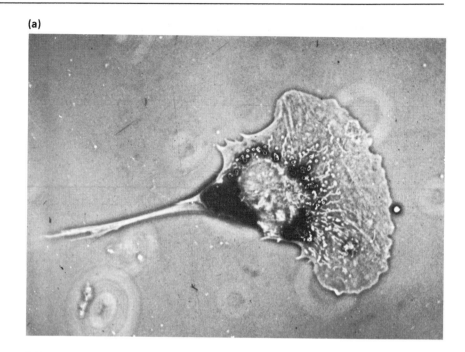

(b)

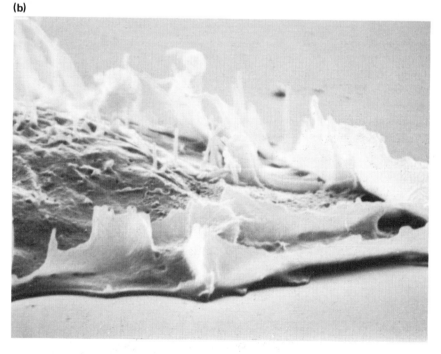

Figure 7.15 (a) Phase contrast micrograph of a well polarised embryonic fibroblast cell. The leading lamellae can be seen. The triangular shape of the cell arises as a consequence of the traction exerted by the leading lamellae and the firmly anchored trailing edge or tail of the cell. The adhesive junction between the substrate and cell tail will eventually break and the rear of the cell will snap back into the main cell body in a way reminiscent

projections which gradually disappear as the cell first makes contact with, and then begins to spread on, the culture vessel surface (Fig. 7.13). The disappearance of these surface projections (often called **filopodia**) is associated with the tendency for thin, sheet-like protrusions of the cytoplasm to extend from the cell body and adhere to the substrate. These protrusions (termed **lamellae**) very often spread to fill the space between adjacent filopodia which are already in contact with the substrate. To begin with, cell spreading occurs radially so that the cell takes on a characteristic 'fried egg' shape, with a central thick zone containing the nucleus being surrounded by an area of thin cytoplasm (Fig. 7.14). After several hours, *the cell loses its radial symmetry and becomes polarised.* Lamellar activity becomes restricted to a more limited area of the cell periphery and the cell can be seen to migrate in the direction of this margin. This region of the cell is called the '**leading lamella**'; it polarises the cell and becomes the advancing margin, and it seems that the activity of the leading lamella pulls the cell along (Figs 7.15a & b).

Electron microscopy has revealed an abundance of actin microfilaments in the cytoplasm of cultured cells. These filaments are very labile, *large bundles of microfilaments suddenly appearing in areas of lamellae spreading or filopodial extension.* Microfilaments are also seen at locations adjacent to the cell-substratum contact. At these sites, a characteristic distribution of actin microfilaments is observed. The filaments appear to originate at, or very close to, the plasma membrane, and run in oblique bands to end up in a less organised meshwork of actin surrounding the nucleus of the cell (Fig. 7.16). It is now accepted that *the mechanical force for locomotion resides in the microfilaments.* In order to exert a mechanical force, the actin filaments must be anchored to the plasma membrane at the sites of cell-substratum adhesion. There is strong evidence to suggest that actin is not directly bound to the cell membrane, but that a specific linking protein known as vinculin connects the terminal actin of a microfilament to an integral component of the membrane (Fig. 7.17).

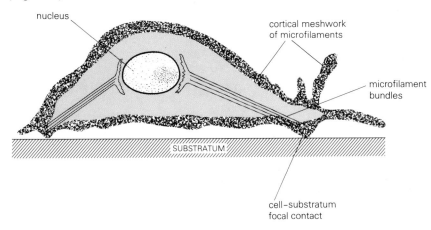

cell–substratum focal contact

Figure 7.16 Schematic diagram of a vertical longitudinal section of a fibroblast in culture to show the distribution of microfilaments in the motile cell. A cortical meshwork of microfibrils also extends into the lamellae while the microfilament bundles normally extend back from the areas of cell–substratum contact to the perinuclear region (not to scale).

Figure 7.15 *continued*

of elastic recoil (courtesy of J. Heath). (b) High-power SEM micrograph of the margin of a fibroblast cell in the region of the leading lamellae. Many projections can be seen, some of which fail to make successful contact with the substrate and are folded back into the cell body (courtesy of J. Heath).

Figure 7.17 Micrograph of a chick fibroblast stained with specific antibodies to vinculin. Visualisation of antibody vinculin using a second antibody and fluorochrome shows localisation of vinculin at sites of cell–substratum adhesion (courtesy of J. Heath).

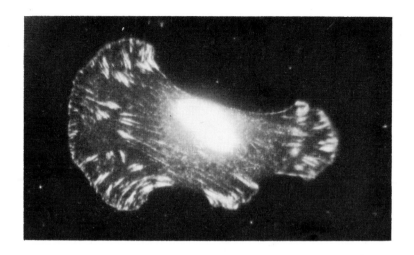

The pulling action required to move a cell is now presumed to come from the arrays of actin filaments described above. The action of small aggregates of myosin present in the cytoplasm will lead to contraction. Since the filaments are fixed to the cell surface at the leading margin of the cell at sites of cell-substrate adhesion, the cell body will be displaced by the myosin-actin contraction towards the leading edge. It is more difficult to explain how the leading lamella is able to extend itself forward prior to making an adhesive contact with the substrate. A form of cytoplasmic streaming caused by an interaction of cortical actin and myosin could force the extension of the lamella in a manner reminiscent of amoeboid locomotion. Other investigators prefer to invoke the controlled polymerisation of actin filaments as a mechanism for extension. The subject of control of cell movement at the molecular level, so that cells can adjust their direction of locomotion in response to environmental cues, remains a most puzzling one.

7.4 Microtubule-based motility

Microtubule structure in all eukaryotic cells is based upon a common plan. Microtubules are about 25 nm in diameter and in section are seen as hollow tubes, with the central lumen being about 15 nm wide. The wall of the microtubule is built up from a circle of sub-units, each 4 nm in diameter and 8 nm long. Normally, 13 of these sub-units go to make up a circle, which can be seen in longitudinal views of microtubules to line up in parallel rows (Fig. 7.18).

The sub-units of microtubules were isolated and purified in quantity from flagella and brain tissue. They were found to consist of a protein complex, named tubulin, with a molecular weight of 110 000 daltons. Two separate polypeptides, called α-tubulin and β-tubulin make up this complex. Despite having a similar molecular weight, approximately 55 000 daltons, the two proteins have distinct, though closely related, amino acid sequences and are thought to have evolved from a single ancestral protein (see §3.2.2 & §3.4).

The two proteins show very little divergence from the lowest to the highest eukaryotes; for example, the β-tubulins of sea urchin flagella and chick brain cells differ in only *one* amino acid. Similarities such as this seen across many phyla suggest that *most mutations disrupt the function of microtubules and are thus lethal and eliminated by selection.*

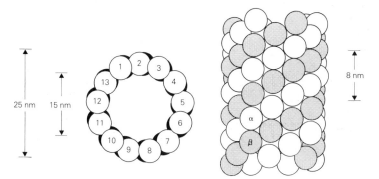

Figure 7.18 Microtubular construction. In cross section the 13 sub-units making up the hollow cylinder are illustrated. The α and β proteins and their helical arrangements are shown in longitudinal section.

The assembly of tubulin molecules into microtubules shares certain features with the assembly of actin into filaments. It is usually associated with the hydrolysis of one molecule of a bound nucleotide for each tubulin, though it is GTP rather than ATP that is used here. There are also a number of other proteins called **MAPS** (microtubule-associated proteins) which form some 5-10 per cent of the protein of microtubules and which promote tubulin polymerisation. Disassembly is also observed under appropriate conditions; indeed the microtubules of non-flagella origin are extremely labile. Exposure to high calcium levels leads to rapid depolymerisation which is just as rapidly followed by microtubule assembly when the calcium is removed.

7.4.1 POLYMERISATION OF TUBULIN

As we have seen, tubulin is a dimer of α-tubulin and β-tubulin and is usually associated with GTP. In addition to this GTP (which is hydrolysed during polymerisation), each tubulin dimer contains a bound molecule of GTP with an unknown function. Microtubule assembly from tubulin dimers seems to occur by a process of self-assembly; given a suitable environment of low calcium concentration, MAPS and levels of free tubulin above a threshold concentration, microtubules can be formed *in vitro*. These seem to share many of the properties of cellular microtubules. Evidence from *in vitro* studies suggests that polymerisation involves two distinct phases, one of **initiation** and the other of **elongation**. The initiation event seems to involve the formation of some multimeric '**nucleating**' centre, following which the addition of more sub-units proceeds rapidly during elongation. In the cytoplasm of eukaryotic cells, nucleating centres can be identified from which microtubules can be seen to grow. These are known as MTOCs (microtubule organising centres) and can be seen in Figure 7.19.

7.4.2 MICROTUBULE ORGANISATION

We have seen earlier that actin microfilaments need to be anchored in the cell for mechanical work to be accomplished in an organised way. Microtubules

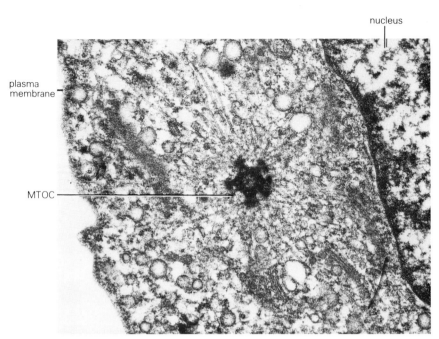

Figure 7.19 Growth (polymerisation) of microtubules from an MTOC found in the cytoplasm of a circulating blood cell. Microtubules can be seen to radiate from the densely stained MTOC.

Cross section

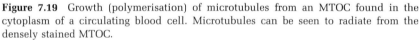

Schematic representation

Figure 7.20 Diagram of centriole structure to show the nine subfibre triplets. Subfibre 1 (or A) is a complete microtubule, while subfibres 2 and 3 (B or C) are incomplete. Other proteins hold the whole array together. Centriolar structure is very similar, if not identical, to that of the basal body of a cilium.

also need to be linked to other components of the cell for their proper function. Most microtubules appear to be anchored at one end; as we shall see later, cilia terminate in basal bodies and cytoplasmic microtubules appear to terminate at nucleating centres (MTOCs) located in regions around the nucleus of the cell. In addition, many microtubules seem to be linked to other cellular structures via the MAPs.

Microtubules in the cytoplasm of a cell in culture can be examined after they have been first depolymerised with specific drugs and then allowed to re-form. The regeneration begins at one or two small structures, called **asters**, and rapidly elongates in the direction of the cell periphery. The *asters are MTOCs, as is the centriole pair present in animal cells*. A **centriole** is composed of nine sets of triplet microtubules, each set containing one complete microtubule fused to two incomplete microtubules. Generally, centrioles are cylindrical in shape (about 0.3 µm long and 0.1 µm wide) and occur in pairs aligned at right angles to each other (Fig. 7.20). Centrioles arise by a process of duplication around the time that DNA synthesis begins in the cell cycle. An existing pair first separates, and a daughter centriole is formed in a plane perpendicular to each original centriole.

Close examination has revealed that most cytoplasmic microtubules do not arise directly from the centriole but from a *densely staining pericentriolar material that surrounds the centriole*. It is now thought that it is this pericentriolar material that is the true MTOC for cytoplasmic microtubules, especially as certain cells which lack centrioles still exhibit the densely stained material. The animal cell centriole has recently been linked to a previously unexpected role: that of *organiser of cell polarity*. It is this role that we shall examine as the first example of microtubule-linked movement.

7.4.3 MICROTUBULES IN CELL LOCOMOTION

In the early stages of cell spreading, when the lamellar cytoplasm is first spreading out in a more or less radial manner from the main cell body, no microtubules are present in the cell periphery. Later, an increasing number of microtubules become evident, extending radially from one or more MTOCs in the nuclear region towards the cell margin. As the cell acquires a polarised morphology, microtubules are commonly found to lie parallel to the long axis of the cell. Exposure to microtubule-disorganising drugs (e.g. colchicine) will have little effect on the initial spreading of cultured cells, but the cells do not subsequently become polarised, and instead of migrating in one direction or another, will appear to try to move in all directions simultaneously. A similar loss of cell polarisation can be seen in growing nerve cells; exposure to such drugs will inhibit the production of the long cell processes characteristic of this cell type. Elongation of the epithelial cells in the developing vertebrate lens is also dependent on polymerisation of microtubules; disrupting drugs will arrest the elongation, leading to the formation of a deformed lens. It seems, then, *that cytoplasmic microtubules are crucial in determining cell polarity*; they are organised by centriole-associated MTOCs as we have seen before, and hence the possibility arises that such *MTOCs 'control' the overall organisation of the cellular skeletal system*.

7.4.4 MICROTUBULES AND CHROMOSOME SEGREGATION

So far, we have examined the state and role of cytoplasmic microfilaments as they apply to interphase cells. The function of microtubules in such cells is still poorly understood compared with what we know of their role in mitosis and meiosis. In many ways, the functions of microtubules as we have discussed them so far seem to emphasise *supportive and controlling elements rather than a strictly motile role*. The best example of a truely motile role for cytoplasmic microtubules is to be found in the function of the **mitotic spindle**.

The mitotic spindle is constructed from microtubules. At the beginning of mitosis, the centrioles of animal cells have divided and moved to the prospective poles of the spindle. Microtubule formation now begins and, depending on the species, spindle structures containing up to 10 000 microtubules are assembled. Typically these are arranged in parallel arrays that converge at the spindle poles, producing a structure that is broad at the midpoint and narrow at either end (Fig. 7.21). It must be made clear at this point that despite the observed correlation between the movement of the

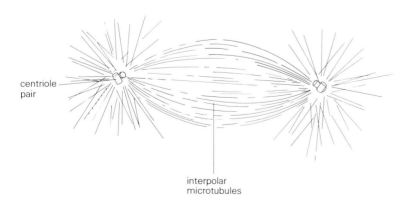

centriole pair

interpolar microtubules

Figure 7.21 The structure of an astral spindle. Experimental removal of centrioles has no effect on the function of the spindle at mitosis though the spindle is broader at the tips than is seen here.

centrioles and the generation of the spindle, *the centrioles do not give rise to the spindle*. Higher plants and some groups of fungi can assemble functional spindles although they lack centrioles. Experimental removal of centrioles from dividing animal cells has no effect on subsequent spindle formation or chromatid separation. It is now believed that, instead of giving rise to spindles, the movement of centrioles is a mechanism which ensures that each daughter cell receives a pair of centrioles during cell division. Thus spindle microtubules divide centrioles just as they do chromosomes.

The true sites acting as centres for spindle formation appear as *aggregates of dense material similar to that described around the centrioles of interphase cells*. They provide fixed nucleation sites for microtubule sub-units which under the correct conditions promote elongation of the microtubules. Other such centres (MTOCs) can be found at the **kinetochores**, specialised attachment regions on the centromeric region of chromosomes. Connection of the chromosomes to the spindle at metaphase establishes two kinds of spindle microtubules that cooperate in anaphase. One type, called **kinetochore microtubules**, runs between the kinetochores and the spindle poles. The second type, called **interpolar microtubules**, extends between the poles without any connection to the chromosomes (Fig. 7.22). Kinetochore microtubules directly link the chromatids with the poles of the spindle, while the interpolar microtubules are normally made up of overlapping, discontinuous segments. This whole microtubule assembly exists in equilibrium with a pool of unpolymerised tubulin sub-units, which can readily convert to the polymerised state or back with changing conditions. The conversion between monomer pool and polymerised microtubules forms at *least part of the force-generating mechanism for chromatid movement at anaphase*.

Movement of the separated chromatids to the spindle poles depends on microtubules and ceases if the spindle is disrupted by anti-microtubule drugs. The movement is very slow compared to cytoplasmic streaming or muscle contraction and is a characteristic of the spindle system rather than the mass of the chromatids. Anaphase movement is complicated; there are two components, one concerned with the actual movement of the chromatids and the second related to the lengthening of the spindle itself. The total

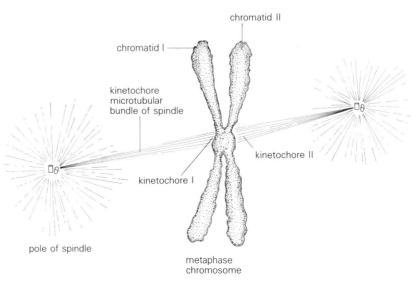

Figure 7.22 The organisation of the kinetochores and their microtubules at mitotic metaphase. Each chromatid has a kinetochore which acts as an MTOC. From these nucleating sites the kinetochore microtubules extend to the opposite poles of the spindle.

chromatid II

chromatid I

kinetochore microtubular bundle of spindle

kinetochore II

kinetochore I

pole of spindle

metaphase chromosome

displacement of the chromatids results from a combination of these two activities and is of great significance to the various theories for anaphase movement.

There are three theories of anaphase movement: the sliding microtubule, the dynamic equilibrium and the sliding microfilament. It is probable that elements of all three mechanisms co-exist in the process of chromatid separation.

7.5 Flagella and cilia

Microtubules are the major skeletal and force-generating elements of flagella and cilia and much of our present understanding of microtubule structure has been obtained from studies on the **axoneme**, the microtubular axis of cilia and flagella. Almost all eukaryotic organisms possess cells that display flagella or cilia at some stage in their life cycle. Only the angiosperms have no flagellated cells of any type, whilst humans have ciliated cells lining the respiratory tract, Fallopian tubes and the ventricles of the brain; flagella are restricted to sperm cells.

The *microtubule systems of cilia and flagella are identical*: whether these organelles are called cilia or flagella is determined largely by their number per cell and their length. When only one or a few are present, they are called flagella; when present in multiples of hundreds, they qualify as cilia. Cilia are usually short (approximately 25 μm); flagella are more than twice as long. Some bacteria also have a locomotory organelle called flagella, but, as we shall see, this is completely different in composition from eukaryotic flagella. Since cilia and eukaryotic flagella are identical, there has been discussion on whether to restrict the term 'flagella' to the prokaryotic structure, leaving the term 'cilia' to mean all eukaryotic flagella as well as cilia. Such a decision would avoid possible confusion but has not yet been agreed as a useful move.

7.5.1 THE AXONEME

The structural arrangement of microtubules in flagella and cilia is uniform (Fig. 7.23). It consists of a circle of nine peripheral double microtubules (the doublets) surrounding two single central microtubules (the central singlets). This arrangement is frequently called the **'9+2' pattern**. Each of the central singlets is a complete microtubule built up from 13 **protofilaments**. Only one of the pair making up a peripheral doublet is complete: this is the **A subtubule** (sometimes called the A subfibre). The second microtubule consists of only 11 protofilaments joined to the side of the A subtubule; this is the **B subtubule**.

The microtubules of the axoneme are held together by connecting elements (Fig. 7.24). The central singlets are connected directly by a bridge and encircled by a sheath. Each doublet is connected to the sheath by a spoke that runs from the A subtubule. The A subtubule also bear two other linking units: **nexin** links connect the A subtubule of one doublet to the B subtubule of the next; a pair of arms also extend from the A subtubule of one doublet to the B subtubule of the next. These arms contain the ATPase enzyme called dynein, which is the *basis of the motive force for movement in eukaryotic flagella and cilia*. All of these connecting elements are repeated at periodic intervals along

Figure 7.23 The microtubule arrangement of the flagellum of the protozoan *Crithidia*. This figure shows the characteristic axoneme structure of the nine peripheral doublets surrounding two single central microtubules. The connecting elements can only be seen as the electrondense diffuse material between the microtubules (courtesy of M. Holiwell).

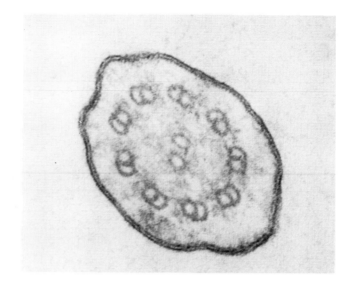

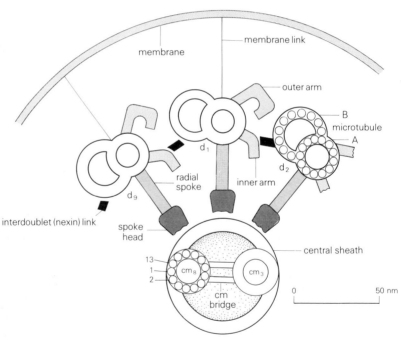

Figure 7.24 Schematic diagram of a section of the axoneme as seen in cross section. Nexin links are shown as connecting the A subtubule of one pair of doublets to the B subtubule of the next. The inner and outer pair of dynein arms on the A subtubule are also shown.

the length of the axoneme and a complete three-dimensional reconstruction of the axoneme has been provided by the work of Amos and Klug.

7.5.2 GENERATION OF MOVEMENT

Flagellar microtubules produce motion by sliding actively over each other (Figs 7.25a & b). Examination of sectioned flagella fixed at different stages of the bending cycle shows that bending occurs without contraction or shortening of the microtubules of the axoneme. It has been shown that

(a)

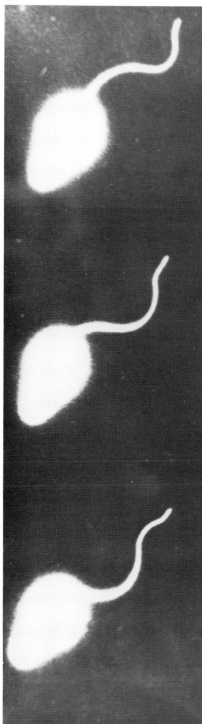
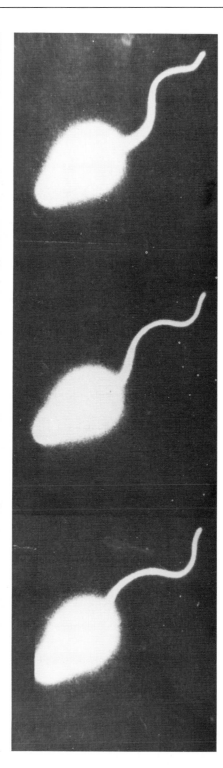

Figure 7.25 The movement of flagellar microtubules. (a) A series of photographs from a high-speed cine film of a protozoan flagellate (*Crithidia oncopatti*). The sinusoidal wave form can be seen clearly only by using this type of photography (courtesy of M. Holwill).

doublet slides over doublet in an active rather than passive manner by an experiment in which the nexin links were enzymatically degraded; addition of ATP caused the microtubules to slide actively out of the axoneme. It seems that the dynein arms produce the force for local sliding of doublets in the axoneme, while the spokes and nexin links hold the doublets together and cause the flagellum to bend in order to accommodate the internal displacement of the doublets during sliding.

(b)

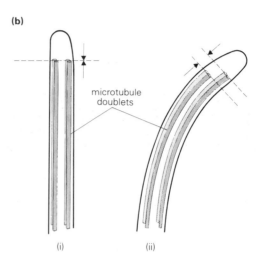

microtubule doublets

(i)   (ii)

Figure 7.25 continued
(b) Schematic diagram to show that flagellar bending occurs without microtubule contraction or shortening. In the unbent condition microtubular doublets extend to within equal distances of the flagellum tip. When bent, doublets on the inside of the bend extend further into the flagellum tip than those on the outside. This is what one would expect to see if sliding occurred between doublets.

The dynein arms provide the force for movement by undergoing cyclic conformational changes while breaking down ATP. Once ATP is bound to dynein, it is hydrolysed. This induces the *conformational change in the dynein arm such that its orientation to a neighbouring microtubule is altered.* At the same instant, a binding site at the tip of the dynein arm is exposed and binds with high affinity to an attachment site on the B subtubule. Attachment releases the imposed shape change in the dynein arm and it moves through a short arc, sliding the B subtubule to a new position some 10 nm distant. The arm is now ready to bind another ATP molecule if this is available in the medium.

7.6 Bacterial flagella

Many bacteria swim by means of thin hair-like appendages called flagella. These are completely unrelated to eukaryotic flagella in structure and mode of generating motion.

Each flagellum is smaller than a single eukaryotic microtubule, being 15-20 nm in diameter. It is built up from several intertwined chains of sub-units of a protein called 'flagellin' which varies from species to species. So far, *none of the various flagellins studies bears any resemblance to either tubulin or actin.* Each flagellum is attached to the bacterium by a basal body consisting of a hook and shaft which anchor the flagellum to the cell wall and plasma membrane (Fig. 7.26). Unlike eukaryotic flagella, the fixed corkscrew or

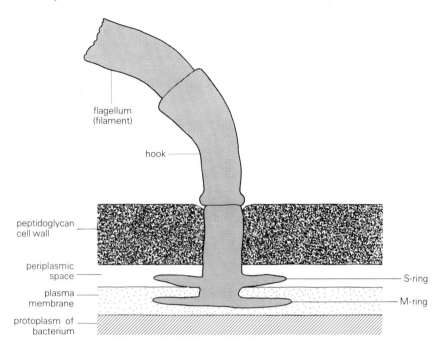

flagellum
(filament)

hook

peptidoglycan
cell wall

periplasmic
space

plasma
membrane

protoplasm of
bacterium

S-ring

M-ring

Figure 7.26 Bacterial flagellum attached to the bacterium by a basal body. In this diagram the arrangement of the basal body as proposed by Berg is shown. The S ring is part of the cell wall and remains stationary. The M ring, which is attached to the base of the flagellum, is moved by a Proton gradient established across the plasma membrane.

helical form of the bacterial flagellum rotates like a propeller! The shaft of the flagellum appears to rotate in its cell wall socket: a rotation driven by a proton gradient across the plasma membrane rather than by ATP breakdown.

7.7   Trends in research

Most recent research in the area of motility has concentrated on the interactions of the cytoskeletal apparatus of cells and co-ordination with other cellular activities. We shall also examine some more detailed biochemistry of the assembly of cytoskeletal elements as this will highlight some fundamental principles of their organisation in cells.

7.7.1   MEMBRANE-CYTOSKELETON INTERACTIONS

One of the most significant concepts to emerge in cell biology in recent years is the idea that cell-surface proteins of various sorts may not be free entities but *linked to the underlying components of the cytoskeleton*. Various lines of evidence point to this possibility even though direct experimental evidence is sparse. A brief description of a few of the more fundamental observations will serve to emphasise the close link between the cytoskeleton and the plasma membrane.

In order to move, a cell must adhere to a substratum against which it can exert traction. Many techniques have been used to show that the adhesions between a cell and its substratum are not distributed uniformly over the lower surface of the cell. Regions of adhesion are confined to a *narrow zone adjacent to the cell margin*, and are only rarely seen beneath the central part of the cell. In a moving cell, the majority of the cell-substratum adhesions are located below the leading lamella, where they are seen as discrete patches of about 1 μm in area which approach to within 30 nm of the substrate. The work of M. Abercrombie and his colleagues showed that these small protrusions from the undersurface of the cell contained filamentous material forming electron-dense '**plaques**' when seen in the electron microscope. Very often these plaques were associated with bundles of microfilaments running obliquely backwards and upwards towards the perinuclear region of the cell. Abercrombie and his associates considered that these regions represented areas of cell-substratum adhesion.

Recent exploitation of an unusual optical microscopial technique (**inter-ference reflection**) first developed by Adam Curtis in the 1960s has confirmed the initial work on fixed cells by allowing examination of the cell-substratum adhesions in living, motile cells. In this technique, cells are illuminated by light passing through the objective of a microscope which also collects the reflected light from the specimen. The thin layer of medium between the cell and the coverslip generates interference patterns in the reflected light, which appear as differences in light intensity to the observer when the separation between coverslip and cell varies. In general, close association between the two surfaces appear darker than more widely separated regions (Fig. 7.27). Two contact sites have now been recognised for cells; **focal contacts** where the gap between cell and substrate is less than 15 nm, and **close contacts** where the separation is nearer 30 nm. A mixture of techniques has now shown that *focal contacts are positioned at the ends of microfilament bundles while close contacts are not*. The latter structures are associated with a loose meshwork of microfilaments which lie just under the plasma membrane.

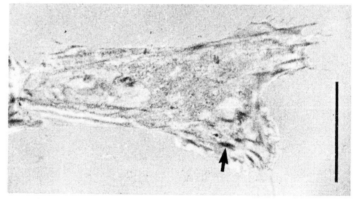

Figure 7.27 Interference reflection micrograph of a chick heart fibroblast. A focal contact is indicated by the arrows. Many others can be seen upon close examination of this print (scale bar 25 μm, courtesy of J. Heath).

A number of techniques have been used to characterise the components of the focal contact. The microfilament bundles which terminate at these sites contain actin, myosin, tropomyosin and other minor elements of the microfilament structure earlier. Another protein, vinculin, is sharply localised in the region of focal contacts and it has been suggested that this protein has a direct role in linking actin filaments to the cell membrane (Fig. 7.17). As yet, there are no known candidates for the actual membrane constituent to which the actin filaments are indirectly anchored, although it is often assumed to be a *transmembrane glycoprotein*.

Fibronectin, a glycoprotein secreted by many cell types, has also been identified in association with the plasma membrane, but on the external face. If fibronectin is added to cells that lack it (such as virus-transformed cells in culture), they undergo morphological changes such that they resemble their non-transformed parent cell. At the same time, actin microfilament bundles appear in the cytoplasm of these cells. This evidence suggests a close structural interaction between cell-surface fibronectin and the microfilament bundles of motile cells. Electron microscopy has provided some support for this, but there is conflicting evidence from other techniques since fibronectin cannot always be found between the cell and substratum at the sites of focal contacts.

Cell surface proteins are induced to congregate into '**patches**' when bound to some molecule (a ligand) in the extracellular environment. The receptors for peptide hormones are an example of this phenomenon in nature, but the same effect can be induced experimentally using a range of ligands. After patches have been induced, they rapidly collect over one pole of a cell in the perinuclear region in a process which requires energy (ATP). The mechanism of capping is unclear but seems to involve the locomotory components of cells, since non-motile cells do not cap. It is as if binding of a suitable ligand to the membrane elicits a response identical to the cell's response when binding to a solid surface; it activates the machinery for locomotion. Two ideas have been proposed as an explanation for capping. One suggests that patches are swept to the area above the nucleus by a continuous flow of membrane in a moving cell. Membrane flow arises in this model as a result of

Figure 7.28 High-power electron micrograph of the cell surface and underlying cytoplasm of a cultured chick embryo fibroblast in the region of an 'arc'. Colloidal gold particles have been used to label a specific surface protein which is then capped (see text). Notice the clumped distribution of the gold over an arc, which is recognised by the high density of microfilaments seen just under the plasma membrane. No gold label is seen over areas without arc microfilaments (courtesy of J. Heath).

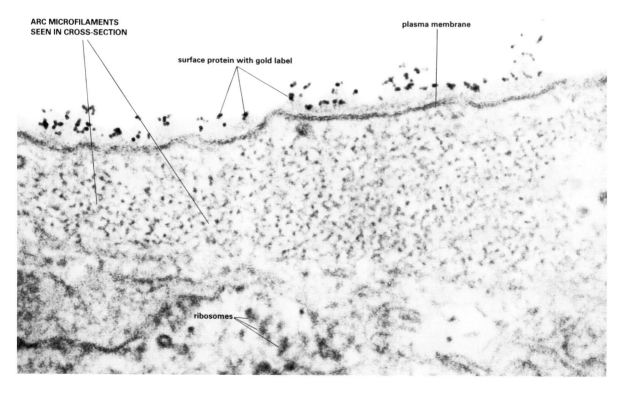

a cycle of membrane endocytosis near the rear of the cell, followed by transportation in the cytoplasm to the leading lamella, where the membrane is re-inserted by exocytosis. While individual membrane proteins can resist membrane flow and remain randomly distributed at the surface, once patched into large aggregates they diffuse too slowly to resist the flow of membrane and hence become capped. The second hypothesis for capping suggests that the protein aggregates interact with the contractile actin microfilaments in the cytoplasm and are actively pulled to the rear of the cell by these microfilaments. There is no clear experimental proof for either of these models, although the demonstration of actin accumulation at regions of a patch does support the latter model. Recently, further experimental evidence favouring the linking of microfilaments to surface proteins patched by ligands came with the demonstration that capping is correlated with the movement of 'arcs' in the microfilament meshwork found just under the cell surface (Fig. 7.28). Arcs are regions of this meshwork that are more densely packed than usual, which arise near, and parallel to, the leading lamella of moving cells and move back before disappearing in front of the cell nucleus.

7.7.2   ASSEMBLY OF ACTIN FILAMENTS AND MICROTUBULES

Cytoplasmic microtubules and actin filaments are assembled from pools of soluble tubulin and actin and disassembled when they are no longer required. The polymerisation of these organelles has a considerable bearing upon their functional organisation and a considerable amount of work has now been done to examine this phenomenon in detail. Polymerisations of both actin and tubulin show similar kinetics; a lag phase due to the formation of an initial fragment (**nucleation**) precedes a more rapid elongation phase. Elongation in both polymers of tubulin and actin is strongly polarised; this is caused by the arrangement of the asymmetric sub-units in a specific orientation in the polymer. In actin filaments and microtubules, the two opposite ends of the polymer assemble and disassemble at very different rates. The fast-growing end of microtubules (the positive end) elongates at over three times the rate of the negative end. The positive end has been shown to be the one that points away from the basal body of a cilium and to which tubulin molecules are added to the growing cilium. For actin filaments, the positive end corresponds to the end that is embedded in the Z-line of striated muscle. In all cells there seem to be proteins that bind selectively to one or the other end of actin filaments and microtubules. These so-called capping proteins must play a major regulatory role in determining the arrangement of the cytoskeleton: allowing for controlled addition and loss of sub-units at either end of the polymer.

References and further reading

Berg, H. C. 1975. How bacteria swim. *Scient. Am.* **233** (Aug.), 36-44.

Cheung, W. Y. 1982. Calmodulin. *Scient. Am.* **246** (June), 48-56.

Dustin, P. 1980. Microtubules. *Scient. Am.* **243** (Feb.), 58-68.

Huxley, H. E. 1965. The mechanism of muscular contraction. *Scient. Am.* **213** (Dec.), 18-27.

Lazarides, E. and J.-P. Revel 1979. The molecular basis of cell movement. *Scient. Am.* **240** (May), 88-101.

Murray, J. M. and A. Weber 1974. The cooperative action of muscle proteins. *Scient. Am.* **230** (Feb.), 58-71.

Porter, K. R. and J. B. Tucker 1981. The ground substance of the living cell. *Scient. Am.* **244** (Mar.), 40-51.

Satir, P. 1974. How cilia move. *Scient. Am.* **231** (Oct.), 44-52.

Glossary

adaptive radiation The evolution of various forms from a single ancestral type.

amniocentesis Removal of a sample of the amniotic fluid surrounding the foetus.

antigen A macromolecule that is capable of stimulating the immune system to produce antibodies.

carbanion Forms when a carbon atom becomes ionised.

coacervate A state formed when two or more differently charged polymeric molecules interact in water.

coenzyme A substance which associates with an enzyme allowing the catalytic process to occur. The molecule may accept or donate an atom or simple group of atoms.

complementation Interaction of two alleles of the same cistron producing a phenotype which is similar to the wild type.

daltons An alternative name for atomic mass unit.

hepatocytes The most common type of cell in the liver.

hydrophilic Having an affinity for water.

hydrophobic Having no affinity for water.

intercalating Insertion of a compound between the adjacent bases in the DNA molecule.

kinetochore The point on the chromosome to which the nuclear spindle attaches.

minimal medium A growth medium which contains a source of carbon (sugar), inorganic nitrogen and mineral salts.

mucopeptide A compound formed from a protein and carbohydrate.

nyctinastic The response in plants to periodic alternations in day and night.

oncogenic Capable of producing cancer.

peptidoglycan A molecule consisting of linear sugar units cross-linked by peptides.

phycobilins Red and blue pigments which are components of the chlorophylls of red and blue-green algae.

progenote An early form of cell, capable of self-replication and a few simple reactions.

prosthetic group A non-protein group which is associated with a protein.

redox pair The alternative oxidised and reduced forms in which a substance can exist.

resonance hybrid   A representation of the structure of a molecule or ion in which delocalisation occurs.

scintillant   A molecule which emits light following an interaction with ionising radiation.

symbionts   Organisms that form an association in which there is mutual benefit to both or all individuals.

template   A molecule which contains information from which other molecules can be synthesised.

transcription   The production of an RNA strand from a DNA **template**.

translation   Synthesis of a polypeptide specified in an mRNA molecule.

transition state   The maximum energy state through which a reaction must pass before it achieves the formation of an intermediate state or product formation.

Index

Numbers printed in *italics* refer to figures while those in **bold type** refer to section numbers.